Lecture Notes in Mathematics
A collection of informal reports and seminars
Edited by A. Dold, Heidelberg and B. Eckmann, Zürich

49

Carl Faith
Rutgers, The State University, New Brunswick, N. J.

Lectures on Injective Modules and Quotient Rings

1967

Springer-Verlag · Berlin · Heidelberg · New York

Library of Congress Catalog Card Number 67-31680. Printed in Germany. Title No. 7369.

TABLE OF CONTENTS

PREFACE TO THE SPRINGER EDITION

These Lectures were written for beginning graduate students, and for that reason are self-contained except for one or two of the chapters at the very end. The main part of the Lectures can be covered in once-weekly sessions of 75-90 minutes each running through two semesters.

I have taken the opportunity to make a number of revisions, have added to the bibliography, but have not included any of the new literature on the subject. Gabriel's thesis listed in the additional bibliography has obtained the theorems of Johnson, Utumi, and Goldie, among others, functorially through the concept of localizing subcategories of abelian categories. Lambek's excellent text, Rings and Modules (Blaisdell 1966), contains some results on quotient rings not covered in these Lectures, and his comments on the literature (Ibid. pp. 166-171) are especially valuable.

SUMMARY OF REVISIONS AND ADDITIONS

(1) I have revised the treatment of the Johnson ring of quotients by first introducing Utumi's ring of quotients, using Lambek's [2] characterization, and obtaining Johnson's ring of quotients as a special case. This appears at the beginning of §8.

(2) I have revised §12 on maximal quotient rings, and also added enough material to get the following theorem:

If R is a prime ring, and if e is an idempotent in the (Johnson) maximal quotient ring S of R such that $K = eSe \cap R \neq 0$, then eSe is the maximal quotient ring of K.

An example is adduced to show that when in addition S is the classical quotient ring of R, then eSe need not be the classical quotient ring of K, and K need not be semiprime. It is an open question whether the theorem stated above can be extended to Utumi's ring of quotients.

Part 3 of §12 has been added. In it, necessary and sufficient conditions are obtained in order that the maximal right quotient ring S of R is also a left quotient ring. If U denotes a minimal right complement ideal of R, these conditions are:

(1) $\Gamma = \mathrm{End}_R\, U$ *is a left Ore domain*.

(2) $US = \Delta U \approx \Delta \otimes_{\Gamma} U$, *where* $\Delta = \mathrm{End}_S\, US$.

Note that since US is a minimal right ideal of S, then Δ is a field.

If S is the classical right quotient ring, then Goldie has shown that (1) is necessary and sufficient for this. Thus in this case we must always have $\Delta U = US$.

(3) To §13 I have added Part 4 of the original paper with S. U. Chase. Let S be a full ring of linear transformations in a vector space over a field K. Let R be a

subring of S such that S is the maximal right quotient ring, and also a left quotient ring, of R . Then there is a "matrix subsystem" $\{D_{ij}\}$ contained in K and a ring isomorphism $R \to \bar{R} \subseteq S$ such that $\bar{R}$ contains all matrices of the form $\{mI + (a_{ij}) \mid m = \text{integer}, a_{ij} \in D_{ij}\}$.

(4) Section 14 is an entirely new section on some implications of Johnson's "transitivity" theorem. If S is the complete ring of linear transformations in a vector space over a field D , and if S is the quotient ring of R , then Johnson's theorem implies that the subring RD generated by R and scalar linear transformations is dense in the finite topology on S . Thus, as in the Faith-Utumi theorem (§9), the quotient ring S may be "obtained" from R by "adding" scalars.

(5) Problems 15-17 have been added.

I wish to thank Fräulein Ruth Schmid for the excellent typing for the Springer Lecture Notes.

Carl Faith
February 14, 1967

INTRODUCTION

A ring S with identity is a classical right quotient ring of a subring T in case: (a) T contains a regular element, that is, an element which is neither a right nor left divisor of zero in T ; (b) each regular $b \in T$ has an inverse $b^{-1} \in S$; and (c) each element $s \in S$ has the form $s = ab^{-1}$, with $a, b \in T$. For brevity, then we say R is a right order in S . The following condition is necessary and sufficient in order that a ring T containing regular elements possesses a classical right quotient ring: if $a, b \in T$ and if b is regular, then there exist $a_1, b_1 \in T$, and b_1 regular, such that $ab_1 = ba_1$. (See either Jacobson [1. p. 118] or Jacobson [2. Appendix B].) If T is a commutative ring with identity, this condition is automatic, since then $b_1 = b$, $a_1 = a$ will satisfy the requirement. If T is an integral domain, not necessarily commutative, this is simply the right Ore condition: $aT \cap bT \neq 0$, for each pair of nonzero $a, b \in T$ (common right multiple property). In this case, T is called a right Ore domain, and its classical right quotient ring is called its right quotient field (Ore [1]).

Now let S be a semisimple ring with identity, that is, let S be a direct sum of minimal right ideals, or equivalently (by the Wedderburn-Artin theorem), let S be a semiprime right Artinian ring with identity. A ring R is said to be right quotient-semisimple in case R has classical right quotient ring S ; R is right quotient-simple in case S is a simple ring.

In a paper in 1958, which made a major contribution to the structure theory for non-commutative rings satisfying the maximum condition (= right Noetherian rings), A.W. Goldie discovered necessary and sufficient conditions on a ring R in order that R be left and right quotient-semisimple. In 1960, he found necessary and sufficient conditions on R in order that R be right quotient-semisimple. The conditions are that R be a semiprime ring, i.e. a ring containing no nilpotent ideals, which satisfies the maximum conditions on annihilator right ideals and complement right ideals. (A complement right ideal is defined to be a right ideal which is maximal with respect to having zero intersection with some right ideal.)

A special case of this theorem (also independently proved by Lesieur and Croisot) states that a right Noetherian prime ring R (e.g. a right Noetherian simple ring) has a classical right quotient ring S which is a full ring D_n of $n \times n$ matrices over a not necessarily commutative field D . In 1964, Faith and Utumi [3] proved that any right order R of D_n contains a subring K_n which is a full ring of $n \times n$ matrices over a right order K in D , and Faith [2] showed that K can be taken to be a simple ring if R is simple. (In either case, K need not contain an identity element, and a change of matrix units in D_n may be required.)

If T has a classical right quotient ring Q , then the natural right T-module Q_T is an essential extension of the right module T_T , that is, Q_T is contained in the injective hull of T_T . Accordingly our development starts in §§1, 2 with the study of injective modules. (Non-unital modules are incorporated in this development.) Baer's criterion for injectivity appears in §1, and the Eckmann-Schopf notion of essential extensions, and the injective hull of a module, are presented in §2.

Quasi-injective modules are introduced in §3, and the double annihilator relation for these modules yield the density theorem for irreducible modules (after Artin [1], Jacobson [2], and Johnson-Wong [1]). The Wedderburn-Artin theorems are well known consequences. In §4, the radical of an arbitrary ring is discussed à la Jacobson. In §5, the Utumi theorems on the structure of the endomorphism ring Λ of an injective module are placed in the setting of quasi-injective modules: The radical $J(\Lambda)$ consists of all endomorphisms whose kernels are essential submodules, and $\Lambda/J(\Lambda)$ is a regular ring. When $J(\Lambda) = 0$, then Λ is a right self-injective ring (cf. Johnson-Wong [2]). Several familiar homological characterizations of Noetherian rings, and of semisimple rings, are given in §6.

In §7, the Findlay-Lambek notion of a rational extension of a module is studied using the approach of Johnson and Wong [2]. The Johnson theorem on the isomorphism between the lattice of (rationally) closed submodules of a module having zero singular submodule, and that of its injective hull is found here.

In 1951, R.E. Johnson defined for a ring R having zero right singular ideal a unique maximal quotient ring R which subsequently (Johnson-Wong [2]) proved to be the injective hull of R_R supplied with a module-preserving ring structure. In 1956, Utumi [1] proved that if R is any ring with identity, then the maximal rational extension $\bar{R}$ of R_R is a ring whose operations preserve the module structure. Unfortunately, Lambek's brilliant remark (Lambek [2]) to the effect that $\bar{R}$ is canonically isomorphic as a ring to $\mathrm{Hom}_\Lambda(E,E)$, where E is the injective hull of R_R , and where $\Lambda = \mathrm{Hom}_R(E,E)$, appeared too late to be included here. Anyway, our program has to be carried through for rings without identity (and this is not just a sophistry - the Utumi-Faith theorem cited above does not seem to succumb to the usual "add an identity" trick); consequently we have the inelegant, unsatisfying development §8 of the Johnson quotient ring $\hat{R}$. Also derived here is the Johnson isomorphism (under contraction) of the respective lattices of closed right ideals of $\hat{R}$ and R . Since $\hat{R}$ is a right self-injective and regular ring, this is the lattice of principal right ideals of $\hat{R}$. (Throughout the text, regular rings are called <u>von Neumann rings</u>; however my nomination for these rings apparently is doomed because I later discovered "von Neumann algebras" elsewhere.)

In §9, we show that R is right quotient-semisimple (resp. quotient-simple) if and only if R is a semiprime (resp. prime) ring, with zero right singular ideal, satisfying the maximum condition on its lattice of closed (= complement) right ideals (and then $\hat{R}$ is the classical right quotient ring of R).

In §10, Goldie's theorems are stated in their original forms. Also included here is Utumi's brilliant short proof of Levitzki's theorem on the nilpotency of nil ideals in Noetherian rings. (Recently Procesi has given an elementary proof of Goldie's theorem for prime rings. An account of this, together with Herstein's reduction to the semiprime case, will appear in Appendix B of Jacobson's revised Colloquium book [2].)

The Faith-Utumi theorem occupies §11, while §12 is a dittoed copy of joint paper with Utumi on various aspects of maximal quotient rings, e.g. the relation between $e\hat{R}e$ and $e\hat{R}e \cap R$, where $e = e^2 \in \hat{R}$.

§13 is a copy of a joint paper with S. U. Chase. Let P denote a direct product of full linear rings, let R be any ring, M an R-module having endomorphism ring S. The following three situations are characterized: (a) $S = P$; (b) $\hat{R} = P$; and (c) $R = P$. Some open problems are cited in §14.

ACKNOWLEDGEMENTS

These notes cover lectures at Pennsylvania State University, Spring 1962, and at Rutgers University Fall and Spring 1962-63. I owe thanks for the encouraging interest of the following auditors: F. Kasch, B. Mueller, R. Rentschler, U. Oberst (at Penn State), and R. Bumby, I. Bentsen, W. Caldwell, R. Cohn, R. Courter, H. Gonshor, R. Heaton, T. MacHenry, B. Osofsky, J. Oppenheim , E. Taft, and S. Weingram (at Rutgers).

During 1961-62 I was fortunate to have opportunities to discuss many of the ideas presented here with Stephen U. Chase. I am grateful to him for these stimulating (and clarifying) conversations, and for devising neat proofs of a number of theorems.

To R. Bumby, W. Caldwell, H. Gonshor, B. Osofsky, and E. Taft, who read proof of various sections, and to Barbara Caldwell and Judith Pittman, who typed, I offer special thanks.

Carl Faith
February 14, 1964

SPECIAL SYMBOLS

§ 0

$\forall$ (resp. $\forall'$); "for all" (resp. "for almost all")
Z,Z; the ring of integers (not to be confused with $Z(M_R)$, the singular submodule of M_R).

§ 1

(on p.8) S^r (resp. S) right (resp. left) annihilator in R of $S \subseteq R$; (cf. S^M)
(on p.12) R_1; the ring obtained from R by freely adjoining an identity.
(on p.12) M_{R_1}; unital module over R_1 obtained from M_R.

§ 2

(B, A, f), $B \geq_f A$, $0 \longrightarrow A \xrightarrow{f} B$ exact; extensions
$B' \geq A$, $B' \geq_f A$, $0 \longrightarrow A \xrightarrow{f} {}'B$ exact; essential extensions
(B, A, f) $=_\theta$ (C,M,h); equivalent extensions
E(M), or $E(M_R)$; the injective hull of M_R.

§ 3

N^R, S^M, S^{MR}, $N^\wedge$, $N^{\wedge M}$.

§ 4

J(R); Jacobson radical of the ring R.
(O:X), (O:M), (I:R)

§ 5

$(S:T) = \{r \in R \mid Tr \subseteq S\}$.
$Z(M_R)$; singular submodule of M_R.
$Z_r(R)$; right singular ideal of R $(= Z(R_R))$.

§ 7

$M \triangledown N$, $(M \triangledown N)_R$: M is an essential extension of N
$M \blacktriangledown N$, $(M \blacktriangledown N)_R$: M is a rational extension of N
C(M), $C(M_R)$; lattice of closed submodule of M_R

§ 8

$\hat{R}$; maximal right quotient ring of R (when $Zr(R) = 0$)

$C_r(R)$; lattice of closed right ideals of R

0. DEFINITIONS

$A \times B$ will denote the cartesian product of two nonempty sets A, B. Below the symbol "$\forall$" (resp. $\forall'$) should be read "for all" (resp. "for almost all").

If R is any ring, not necessarily containing an identity element, and if $(M,+)$ is an additive abelian group, then M is a right R-module in case there exists a mapping $(x,r) \to xr$, $x \in M$, $r \in R$, of $M \times R$ into M satisfying the following conditions:

$$\begin{aligned} &(1) \quad (x+y)r = xr + yr \\ &(2) \quad x(r+s) = xr + xs \qquad\qquad (\forall\ x, y \in M,\ r, s \in R) \\ &(3) \quad x(rs) = (xr)s \end{aligned}$$

M_R denotes the fact that M is a right R-module; a left R-module $_RM$ is defined symmetrically; a submodule of M_R is a subgroup N of M such that $xr \in N$ $\forall$ $x \in N$, $r \in R$.

M_R is a unital module in case R contains an identity element 1, and $x1 = x$ $\forall$ $x \in M$.

N_R is homomorphic to M_R in case the additive group $(N,+)$ is homomorphic to $(M,+)$ under a mapping $\varphi: x \to x'$, $x \in N$, $x' \in M$, satisfying $(xr)' = x'r$ $\forall$ $x \in N$, $r \in R$. Then we write $N_R \sim M_R$. The kernel $Ker(\varphi)$ of φ is a submodule of M_R; φ is an isomorphism in case $Ker(\varphi) = 0$, and we write $N_R \cong {}_{\varphi}M_R$.

If K_R is any submodule of M_R, then the difference group $M - K$ is an R-module under the definition $(m + K)r = mr + K$ for any coset $m + K \in M - K$, and any $r \in R$; the natural homomorphism of the group M on $M - K$ is also a module homomorphism, that is, $M_R \sim (M - K)_R$.

If F is a nonempty collection of sets, then F is indexed by a set I in case to each $i \in I$ there corresponds a set $X_i \in F$ and the mapping $i \to X_i$ is onto F. The notation $\{X_i \mid i \in I\}$ will always denote a family of sets (possibly each set consisting of a single element) indexed by a set I; $\cup_{i \in I} X_i$ denotes the set theoretical union of a family $\{X_i \mid i \in I\}$ of sets.

If $\{X_i \mid i \in I\}$ is any family of sets, the totality $\times_{i \in I} X_i$ of mappings x of I into $\cup_{i \in I} X_i$ such that $x(i) \in X_i$ $\forall$ $i \in I$ is called the cartesian product of $\{X_i \mid i \in I\}$. Instead of $x(i)$ one usually writes x_i, and this is called the i-th coordinate of x; x itself is usually represented as: $x = (\ldots, x_i, \ldots)$. The mapping $\pi_i: x \to x_i$, $x \in \times_{i \in I} X_i$ is called the projection of $\times_{i \in I} X_i$ on X_i (or the projection of $\times_{i \in I} X_i$ on its i-axis).

If $\{M_i \mid i \in I\}$ is any family of R-modules, then the direct product $\Pi_{i \in I} M_i$ is the cartesian product $\times_{i \in I} M_i$ which is an R-module under the operation:

$$xr = (\dots, x_i r, \dots) \quad \forall \quad x = (\dots, x_i, \dots) \in X_{i\in I} M_i \ , \ r \in R \ .$$

The <u>direct sum</u> $\sum_{i\in I} \oplus M_i$ is the submodule of $\prod_{i\in I} M_i$ consisting of all $x = (\dots, x_i, \dots)$ such that $x_i = 0 \ \forall' \ i \in I$.

Let $\{N_j \mid j \in J\}$ be any nonempty collection of submodules of a module M_R . Then

$$\sum_{j\in J} N_j = \{\sum_{j\in J} x_j \mid x_j \in N_j \ , \ x_j = 0 \ \forall' \ j \in J\}$$

is the smallest submodule of M_R containing $N_j \ \forall \ j \in J$, and is called the <u>sum of the</u> $\{N_j \mid j \in J\}$. The family $\{N_j \mid j \in J\}$ of submodules is <u>independent</u> in case

$$N_i \cap \sum_{i\neq j\in J} N_j = 0 \qquad \forall \ i \in J \ .$$

Then the natural homomorphism $\eta : (\dots, x_i, \dots) \to \sum_{i\in I} x_i$ of $\sum \oplus_{j\in J} N_j$ onto $\sum_{j\in J} N_j$ is an isomorphism. The converse also is easily proved: If the natural homomorphism of $\sum \oplus_{j\in J} N_j$ onto $\sum_{j\in J} N_j$ is an isomorphism, then the family $\{N_j \mid j \in J\}$ is independent. The condition below is also equivalent to independence of $\{N_j\}$:

If $x = \sum_{j\in J} x_j \in \sum_{j\in J} N_j$ (here $x_j = 0 \ \forall' \ j \in J$), then $x = 0$ if and only if $x_j = 0 \ \forall \ j$.

If M and N are additive abelian groups, Hom (M,N) will denote the set of all homomorphisms of M into N; Hom (M,N) is itself an additive abelian group with respect to addition $f + g$ of mappings $f,g \in$ Hom (M,N) defined by:

$$(f + g)(x) = f(x) + g(x) \qquad \text{for all} \quad x \in M \ .$$

If in addition M_R and N_R are R-modules, then

$$\text{Hom}_R(M,N) = \{f \in \text{Hom}\ (M,N) \mid f(x)r = f(xr) \quad \forall \quad x \in M \ , \ r \in R\} \ .$$

If M is a right R-module and a left S-module, and if $(sx)r = s(xr) \ \forall \ s \in S, \ r \in R$, then M is an (S,R)-<u>bimodule</u>, notationally ${}_S M_R$.

If M is any additive abelian group, then $E =$ Hom (M,M) is a ring, called the endomorphism ring of the abelian group M , under the operations of addition $f + g$ and composition $f \circ g$ of mappings $f,g \in E$. Here

$$(f \circ g)(x) = f(g\ (x)) \qquad \forall \ x \in M, \ f,g \in E \ .$$

Then, M is a left E-module under the definition

$$fx = f(x) \quad \forall \quad f \in E \ , \quad x \in M \ .$$

If, in addition, M is a right R-module, then $S = \mathrm{Hom}_R(M,M)$ is a subring of E, called the <u>endomorphism ring of the module</u> M_R, and then M is an (S,R)-bimodule.

<u>EXAMPLES</u>

1. Any additive abelian group G is both a right and left Z-module; in fact G is a (Z,Z)-bimodule. Here, as throughout, we adopt the Bourbaki notation Z to indicate the ring of integers.

2. If R is any ring, then any right ideal I is a right R-module, I_R, and any ideal P of R is an (R,R)-bimodule; in particular, R itself is a right (resp. left) module over R, denoted by R_R (resp. ${}_RR$). If e is any idempotent of R (that is, any element of R satisfying $e = e^2$), then eR is a right ideal which is also a left eRe-module. Furthermore,

$$\mathrm{Hom}_R(eR,eR) \approx eRe$$

under the correspondence

$$f \longrightarrow f(e), \qquad \forall \quad f \in \mathrm{Hom}_R(eR,eR).$$

(Hint: Show that $f(e) \in eRe$, and that $f(x) = f(e)x \quad \forall \quad x \in eR$.)
If R is any ring with identity element, conclude that $\mathrm{Hom}_R(R,R) \cong R$.

3. If K is any division ring (that is, a not necessarily commutative field), then a right vector V space over K is a unital right K-module (and conversely). In this case the endomorphism ring $\mathrm{Hom}_K(V,V)$ of the module V_K is called the <u>full linear ring in the vector space</u> V <u>over</u> K. We shall adopt the following terminology: A <u>right</u> (resp. left) <u>full</u> (linear) <u>ring</u> R is a ring isomorphic to $\mathrm{Hom}_K(V,V)$ for some right (resp. left) vector space V over K.

Let A_R, B_R be modules. If f is a homomorphism of A into B, the range (or image) of f is denoted by $\mathrm{Im}(f)$. Then, f is an <u>epimorphism</u> in case $B = \mathrm{Im}(f)$, and f is a <u>monomorphism</u> in case $\mathrm{Ker}(f) = 0$. Thus, in this language, a homomorphism $f: A \to B$ is an isomorphism of A and B if and only if f is both an epimorphism and a monomorphism. "Maps" and "homomorphism" will be used synonymously.

A sequence of homomorphisms (or maps)

$$A_m \longrightarrow A_{m+1} \longrightarrow \ \cdots \ \longrightarrow A_n \qquad\qquad m+1 < n$$

is said to be exact if for each $m < q < n$ we have $\mathrm{Im}(A_{q-1} \to A_q) = \mathrm{Ker}(A_q \to A_{q+1})$. Thus, $A \to B$ is a monomorphism if and only if $0 \to A \to B$ is exact, and $A \to B$ is an epimorphism if and only if $A \to B \to 0$ is exact.

In particular, if the sequence

$$0 \longrightarrow A \longrightarrow B \longrightarrow C \longrightarrow 0$$

is exact, then $A \to B$ is a monomorphism, and $B \to C$ is an epimorphism with kernel $= \mathrm{Im}(A \to B)$. Thus, the sequence above may be replaced by

$$0 \longrightarrow A \longrightarrow B \longrightarrow \mathrm{Im}(A \to B)$$

If A is a submodule of B, then there is a natural epimorphism $B \to B - A$, and the sequence

$$0 \longrightarrow A \longrightarrow B \longrightarrow B - A$$

is exact (where $A \to B$ is the identity map).

An exact sequence $0 \longrightarrow A \xrightarrow{f} B$ <u>splits</u> if $f(A)$ is a direct summand of B. In this case, the diagram (1), where i is the identity map,

$$(1) \qquad \begin{array}{ccccc} 0 & \longrightarrow & A & \xrightarrow{f} & B \\ & & i\uparrow & & \\ & & A & & \end{array} \qquad\qquad (2) \qquad \begin{array}{ccccc} 0 & \longrightarrow & A & \xrightarrow{f} & B \\ & & i\uparrow & \swarrow g & \\ & & A & & \end{array}$$

can be embedded in a commutative diagram (2). To see this, write $B = f(A) \oplus K$, where K is some submodule of B, and then the projection g of B on $f(A)$ with respect to K has the desired property. Conversely, if each f is any monomorphism $0 \longrightarrow A \xrightarrow{f} B$ such that the diagram (1) can be completed as in (2), then $f(A)$ is a direct summand of B. To see this, note that $gf(a) = a \ \forall\, a \in A$, consequently $fg\, f(a) = f(a) \ \forall\, a \in A$, i.e., $fg(y) = y$ for all $y \in f(A)$. Hence, letting $\varphi = fg$, one easily sees that $B = f(A) \oplus \mathrm{Ker}(\varphi)$.

Let F_R be a unital module and X a subset of F_R. We say that F_R is a <u>free</u> module with <u>free basis</u> X if and only if:

(1) $F = \sum_{x \in X} xR$; and

(2) $\{xR \mid x \in X\}$ is an independent family of submodules of F_R.

It is easy to see that (1) and (2) are equivalent to

(3) each $y \in F$ can be written in one and only one way as a finite sum

$$y = \sum x_i r_i \qquad (x_i \in X \ , \ r_i \in R \ ).$$

If X is any set we consider the family $\{R_x \mid x \in X\}$ of rings, where R_x is a ring isomorphic to R, and define

$$F(X) = \sum_{x \in X} \oplus R_x \ .$$

Let 1_x denote the identity of R_x. Then, if we identify $x \in X$ with the element

$$(\ldots,0,\ldots,1_x,\ldots,0,\ldots) \in F(X)$$

we see that $F(X)$ is a <u>free</u> module <u>with a free basis</u> X.

1. INJECTIVE MODULES

1. **DEFINITION.** A module M_R is injective in case each pair of modules B_R, A_R, with $B_R \supseteq A_R$, has the property: each $f \in \mathrm{Hom}_R(A_R, M_R)$ can be extended to (or is induced by) an element of $\mathrm{Hom}_R(B_R, M_R)$.

We shall show that Definition 1 is equivalent to:

1'. **DEFINITION.** A module M_R is injective in case any row exact diagram (1) can be embedded in a commutative diagram (2).

(1) 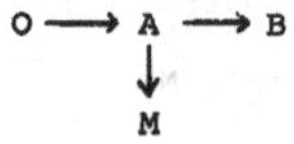

(2) 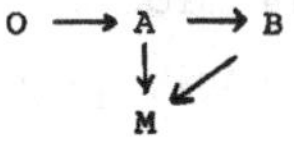

Certainly any module which is injective in the sense of 1' is injective in the sense of 1, since if $A_R \subseteq B_R$, then the injection map $i: A \to A$ is a monomorphism of A in B. Then, by 1', any diagram below can be completed by a map denoted by $- - - - \to$.

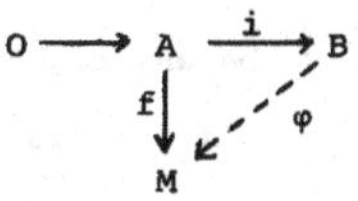

Thus, φ extends the homomorphism $f: A \to M$, and so 1 holds.

Conversely, let M_R be injective in the sense of 1, and consider the row-exact diagram below on the left.

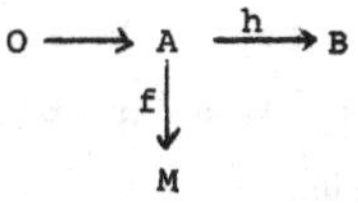

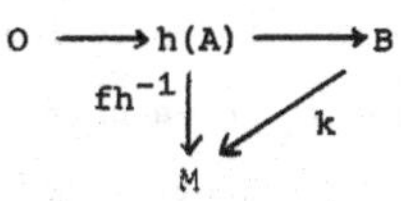

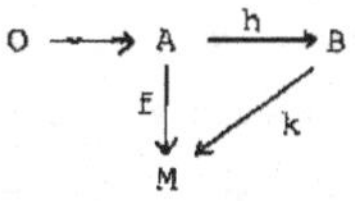

Then $h(A) \subseteq B$ and $fh^{-1}: h(A) \to M$. Hence, by 1, fh^{-1} can be extended to a map k of B in M (center diagram). Thus, the diagram on the left can be completed as indicated on the right and therefore 1' holds.

Because of the simplicity of unital modules, it is convenient to consider a restricted notion of injectivity: If R is a ring with identity, then a unital module M_R is u-injective in case each unital module B_R has the following property: If A_R is a submodule of B_R, then any $f \in \mathrm{Hom}_R(A_R, M_R)$ can be induced by an element $\in \mathrm{Hom}_R(B_R, M_R)$.

Clearly, any injective unital module H_R is u-injective. Conversely, if M_R is u-injective, and if G_R is any R-module, we can write $G = G^1 \oplus G^0$, where $G^1 = G1$, and $G^0 = \{x \in G \mid x1 = 0\}$ are submodules. Then, if H is any submodule, $H = H^1 \oplus H^0$. Since H^1, G^1 are unital modules, if $h \in \mathrm{Hom}_R(H,M)$, the restriction h_1 of h to H_1 has an extension $g_1 \in \mathrm{Hom}_R(G^1, M)$. Now define $g \in \mathrm{Hom}_R(G,M)$ by

$$g(x_1 + x_0) = g_1(x_1) , \qquad x_1 + x_0 \in G , \quad x_1 \in G^1, \quad x_0 \in G^0 .$$

Then, g is the desired extension of h. We have proved:

2. **PROPOSITION.** Let M_R be a unital module. Then M_R is u-injective if and only if M_R is injective.

EXAMPLE. Let R denote a field, and let V be any right vector space over R, i.e. V_R is any unital module. Then by Zorn's Lemma if A_R is a vector subspace of a vector space B_R, any vector basis of A_R can be extended to a vector basis of B_R, that is, A_R is a direct summand of B_R, $B_R = A_R \oplus C_R$. It follows that any homomorphism f of A_R into V_R can be extended to a homomorphism g of B_R into V_R, e.g., let $g = f$ on A_R and $g = 0$ on C_R. Since V_R is therefore u-injective, it is injective.

The proof that V_R is injective has the following corollary: Let R be a ring with identity such that each submodule A_R of a unital module B_R is a direct summand. Then each unital module M_R is injective. We shall encounter a class of rings having this property, more general than the class of fields, in a later section.

This next proposition will be useful in applications of the notion of injective modules.

3. PROPOSITION. If $M_R \approx_g N_R$, then M_R is injective if and only if N_R is.

PROOF. Consider any diagram

$$\begin{array}{ccccc} 0 & \longrightarrow & A & \longrightarrow & B \\ & & f\downarrow & & \\ & & M & & \end{array}$$

in which the row is exact, and assume that N_R is injective. Since N is injective, we can embed the diagram (on the left)

$$\begin{array}{ccccc} 0 & \longrightarrow & A & \longrightarrow & B \\ & & gf\downarrow & & \\ & & N & & \end{array} \qquad\qquad \begin{array}{ccccc} 0 & \longrightarrow & A & \longrightarrow & B \\ & & gf\downarrow & \swarrow h & \\ & & N & & \end{array}$$

into a commutative diagram (on the right). Then the diagram below is commutative (and it follows that M_R is injective):

$$\begin{array}{ccccc} 0 & \longrightarrow & A & \longrightarrow & B \\ & & f\downarrow & \swarrow g^{-1}h & \\ & & M & & \end{array}$$

REMARKS

4. Each direct summand of an injective (resp. u-injective) module is injective (resp. u-injective).

PROOF. Let M_R be injective, write $M_R = P_R \oplus Q_R$, and consider the row-exact diagram (on the left):

$$\begin{array}{ccccc} 0 & \longrightarrow & B & \longrightarrow & A \\ & & f\downarrow & & \\ & & P & & \end{array} \qquad \begin{array}{ccccc} 0 & \longrightarrow & B & \longrightarrow & A \\ & & f\downarrow & \swarrow g & \\ & & M & & \end{array} \qquad \begin{array}{ccccc} 0 & \longrightarrow & B & \longrightarrow & A \\ & & f\downarrow & \swarrow \pi g & \\ & & P & & \end{array}$$

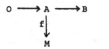

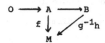

Since M_R is injective, there exists a homomorphism $g\colon A \to M$ such that the center diagram above is commutative. But if π denotes the projection of M on P with respect to Q , then the diagram on the right is commutative, and P_R is therefore injective. (If M_R is u-injective, then M_R is injective by 2, and P_R is injective by the result just proved. Then P_R is u-injective.)

5. A direct product of a family $\{M_i \mid i \in I\}$ of modules is injective (resp. u-injective) if and only if M_i is injective (resp. u-injective) $\forall\, i \in I$.

PROOF. Since M_i is a direct summand of $M = \prod_{i \in I} M_i$, injectivity (resp. u-injectivity) of M implies that of M_i $\forall\, i$, by Remark 4.

Conversely, let M_i be injective $\forall\, i \in I$, and consider a row-exact diagram on the left:

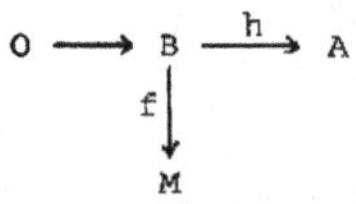

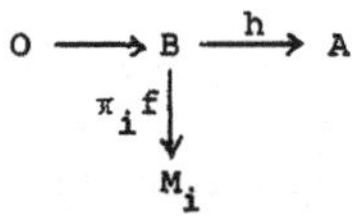

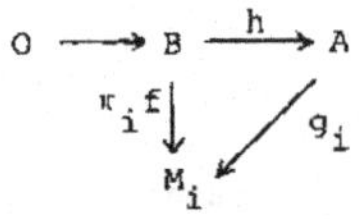

If π_i is the projection of M on M_i , then we obtain a diagram in the center which by injectivity of M_i can be embedded in a commutative diagram on the right (above). We now define a map (= homomorphism) $g: A \to M$ as follows:

$$g(a) = (\ldots, g_i(a), \ldots)$$

Furthermore the diagram

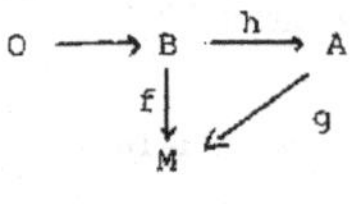

is commutative, since if $b \in B$, then

$$\begin{aligned} gh(b) &= (\ldots, g_i h(b), \ldots) \\ &= (\ldots, \pi_i f(b), \ldots) = f(b) \;, \end{aligned}$$

i.e., $f = gh$ as required. (If M_i is u-injective, then M_i is injective, and the result for u-injective modules follows from the one for injective modules.)

We note that the notion of injectivity of a module M_R is defined "externally" to the module M_R itself. We will show that in many cases, notably for rings considered as modules over themselves, that an "internal" criterion for injectivity can be given. Although no such "internal" criterion for injectivity of arbitrary modules is known, we show below that for u-injective modules a "test module" always exists. A module B_R is a test module for the class $\{R\}$ of all modules over R in case the following equivalence holds:

A module $M_R \in \{R\}$ is injective if and only if each submodule A_R of B_R has the property that each $f \in \text{Hom}_R(A,M)$ can be induced by an element of $\text{Hom}_R(B,M)$.

The proposition below shows that for a ring R with identity a test module always exists, and is in a certain sense the simplest possible, namely R itself. This result is an immediate consequence of the theorem below, which was proved by R. Baer [1].

6. THEOREM. (R. Baer). Let R be a ring with identity, and let M_R be any unital module. Then M_R is injective if and only if for each right ideal I of R and each $f \in \text{Hom}_R(I,M)$ there exists $m \in M$ such that

$$f(x) = mx \quad \forall \quad x \in I .$$

PROOF. Necessity. Assume that M_R is injective, and let $f \in \text{Hom}_R(I,M)$, where I is any right ideal of R. Then f has an extension $g \in \text{Hom}_R(R,M)$, and $m = g(1)$ has the desired property.

Sufficiency. By 2, it suffices to consider unital modules A_R, B_R with $A_R \subseteq B_R$, and a mapping $f \in \text{Hom}_R(A,M)$. Let P denote the collection of all pairs (A',f'), where A' is a submodule of B containing A, and where $f' \in \text{Hom}_R(A',M)$ extends f. We partially order P by decreeing that $(A',f') \geqq (A'',f'')$ if and only if $A' \geq A''$ and f' induces f''. If $\{(A_i,f_i) \mid i \in I\}$ is any linearly ordered subset of P, then $A_0 = \cup_{i \in I} A_i$ is a submodule of $B \supseteq A$. We now define a homomorphism $f_0: A_0 \to M$ as follows: If $a \in A$, then $a \in A_i$ for some $i \in I$, and then we let $f_0(a) = f_i(a)$. Since $f_j(a) = f_i(a)$ for all $j \geq i$, $f_0(a)$ is defined, and $(A_0,f_0) \in P$. By Zorn's lemma, P contains a maximal element, which we denote by (A_0,f_0). If $A_0 = B$, we are finished, since then f_0 is the desired extension to B of f. We shall show that $A_0 \neq B$ leads to a contradiction. Hence assume there exists $x \in B$, $x \notin A_0$. Then $C = A_0 + xR$ is a submodule of B properly containing A_0 . Now $I = \{r \in R \mid xr \in A_0\}$ is a right ideal of R , and the correspondence

$$\theta: r \longrightarrow f_0(xr) \qquad (r \in I)$$

is an element of $\mathrm{Hom}_R(I,M)$. By hypothesis, there exists $m \in M$ such that $\theta(r) = f_0(xr) = mr \ \forall \ r \in I$. Consider the correspondence

$$f_0': a + xr \longrightarrow f_0(a) + mr \qquad (a \in A_0,\ r \in R).$$

If $a + xr = b + xt \in C = A_0 + xR$, $a,b \in A_0, r,t \in R$, then $x(r - t) = b - a \in A_0$, so that $r - t \in I$, and therefore

$$f_0[x(r - t)] = m(r - t) = mr - mt .$$

But

$$f_0[x(r - t)] = f_0(b - a) = f_0(b) - f_0(a) ,$$

so that

$$f_0(a) = mr = f_0(b) + mt ,$$

and f_0' is therefore an element of $\mathrm{Hom}_R(C,M)$. Since f_0' extends f_0 , we have $(C,f_0') > (A_0,f_0)$, which is the desired contradiction.

COROLLARY. If R is a ring with identity, then R_R is a test module for $\{R\}$.

PROOF. A module M_R is injective if and only if each right ideal I of R has the property stated in 6. If $f \in \mathrm{Hom}_R(I,M)$ and if $f(x) = mx \ \forall \ x \in I$, then the mapping $g \in \mathrm{Hom}_R(R,M)$ defined by $g(y) = my \ \forall \ y \in R$ extends f . Conversely, if $g \in \mathrm{Hom}_R(R,M)$ extends f , then $f(x) = mx \ \forall \ x \in I$, where $m = g(1)$. This proves that R is a test module for $\{R\}$.

For easy reference, we say that an arbitrary module M_R over an arbitrary ring R satisfies <u>Baer's condition</u> in case for each map $f: I \to M$ of a right ideal I of R into M_R there exists an element $m \in M$ such that $f(x) = mx \ \forall \ x \in I$. In addition, with 6 in mind, we say that <u>Baer's criterion holds for a class</u> C <u>of modules over</u> R in case $M_R \in C$ is injective if and only if M_R satisfies Baer's condition. In this language, 6 states that Baer's criterion holds for the class of unital modules over any ring with identity.

Following Baer [1], if I is a right ideal of a ring R , then a module M_R is I-complete in case there exists for every map $f: I \to M_R$ an element $m \in M$ such that $f(x) = mx \ \forall \ x \in I$. For each non-empty subset S of R , $S^r = \{x \in R \mid Sx = 0\}$ is a right ideal of R , and $S^M = \{y \in M \mid yS = 0\}$.

7. PROPOSITION. If R is a ring with identity, if M_R is a unital module, and if $I = xR$, where $x \in R$, then M_R is I-complete if and only if

$$(x^r)^M \subseteq Mx .$$

PROOF. First assume that M is I-complete, let $y \in (x^r)^M$, and consider the correspondence $f: xr \to yr \ \forall \ r \in R$. If $xr = xr'$, $r, r' \in R$, then $r - r' \in x^r$, so that $y(r - r') = 0$, and $yr = yr'$. Thus, f is a map of I_R in M_R . Consequently, there exists $m \in M$ such that $f(w) = mw \ \forall \ w \in I$, in particular, $y = f(x) = mx$. Thus, $(x^r)^M \subseteq Mx$.

Conversely, assume that $(x^r)^M \subseteq Mx$. If $f: I \to M$ is any map of I in M , where $I = xR$, then

$$f(x)y = f(xy) = f(0) = 0 \qquad \forall \ y \in x^r ,$$

so that $f(x) \in (x^r)^M$. Consequently, $f(x) = mx$ for some $m \in M$, and clearly $f(w) = mw \ \forall \ w \in I = xR$. Thus, M is I-complete.

An element x is a left divisor of zero in R in case there exists $0 \neq y \in R$ such that $xy = 0$, that is , in case $x^r \neq 0$; R is an integral domain in case $R \neq 0$, and $x^r = 0 \ \forall \ x \in R$.

Now if $x^r = 0$ in R , and if M_R is unital, then $(x^r)^M = M$, so that $Mx = M$ is a necessary and sufficient condition for xR-completeness of M_R . If R is an integral domain, then M is complete with respect to each principal ideal $I = xR$, $x \in R$, if and only if $Mx = M \ \forall \ x \in R$. A module with this latter property is said to be <u>divisible</u>, and we have proved:

8. COROLLARY. Let R be an integral domain with identity element, and let M_R be a unital module. Then:

(1) If M_R is injective, then M_R is divisible;

(2) If M_R is divisible, then M_R is xR-complete $\forall\, x \in R$.

Thus, if R is a principal right ideal domain, that is, if R is an integral domain such that each right ideal is principal we have:

9. COROLLARY. If R is a principal right ideal domain, and if M_R is any unital module, then M_R is injective if and only if it is divisible.

An additive abelian group G is said to be <u>divisible</u> in case G_Z is a divisible module. Since Z is a principal ideal domain, by 9, we know that G is divisible if and only if G_Z is injective.

The main aim of this section is to show that any module M_R can be embedded in an injective module, that is, that there exists an injective module N_R and a monomorphism $\varphi\colon M_R \to N_R$. As we show later in this section it is enough to do this for unital modules. This first was done by R. Baer [1]. Later, Eckmann and Schopf showed that the result for unital modules was a consequence of the corresponding result for additive abelian groups. Since their method eliminates use of the theory of ordinal and cardinal numbers, we adopt it accordingly. However, Baer's construction is in many ways instructive.

10. LEMMA. If G is an additive abelian group then there exists a divisible group D and a monomorphism $\varphi\colon G \to D$.

PROOF. Let $\{g'\lambda \mid \lambda \in \Lambda\}$ be a system of generators of G. Then if $g \in G$, there exists a finite subset $\lambda_1, \dots, \lambda_n \in \Lambda$, and integers $z_1, \dots, z_n \in Z$ such that $g' = \Sigma_1^n z_i g'_{\lambda i}$. Now let $I_\lambda = (g_\lambda)$ be the infinite cyclic group with generator g_λ, $\lambda \in \Lambda$. Then the general element $g \in F = \sum_{\lambda\in\Lambda} \oplus I_\lambda$ has the form

$$g = \sum_{\lambda\in\Lambda} z_\lambda g_\lambda , \qquad z_\lambda \in Z$$

where all but a finite number of the integers z_λ are equal to zero. Consider the correspondence

$$f: g \to g' = \sum_{\lambda\in\Lambda} z_\lambda g'_\lambda , \qquad g \in F , \quad g' \in G .$$

If $g = 0$, then $z_\lambda = 0$ for all $\lambda \in \Lambda$, so that $g' = 0$, whence f is a map, and G is canonically isomorphic to $F - K$, $G \cong \varphi\, F - K$, where $K = Ker(f)$. Considering I_λ as a subgroup of the additive group Q_λ of rational numbers, F is a subgroup of the divisible group $H = \Sigma_{\lambda\in\Lambda} \oplus Q_\lambda$. Now the homomorphic image of a divisible group is again divisible, whence $D = H - K$ is a divisible group containing $F - K$, and $\varphi: G \to F - K$ is the desired monomorphism.

11. **PROPOSITION.** If G is any divisible additive abelian group, and if R is any ring with identity element, then $F = Hom_Z(R,G)$ is an injective R-module.

PROOF. F is a unital R-module under an operation $\varphi \circ r$, $\varphi \in F$, $r \in R$, defined by

$$(\varphi \circ r)[x] = \varphi[rx] , \qquad x \in R .$$

Now let A_R, B_R be R-modules, $A \supseteq B$, and let $f \in Hom_R(B,F)$. If $r \in R$, $b \in B$, then

$$f(br) = f(b) \circ r ,$$

so that

$$f(br)[x] = (f(b) \circ r)[x] = f(b)[rx], \quad \forall\ x, r \in R, b \in B .$$

Now let ψ denote the correspondence:

$$\psi: b \to f(b)(1) , \qquad \forall\ b \in B .$$

Clearly $\psi \in Hom_Z(B,G)$, and, since G_Z is injective, ψ has an extension $\bar{\psi} \in Hom_Z(A,G)$.

Now let $\bar{f}: a \to \bar{f}(a)$ denote the mapping of A into F defined by

$$\bar{f}(a)[x] = \bar{\psi}(ax) , \qquad \forall\ a \in A,\ x \in R .$$

If $r \in R$, then

$$\bar{f}(ar)[x] = \bar{\psi}(ar \circ x) ,$$

also

$$\bar{f}(a)[rx] = \bar{\psi}(ar \circ x) .$$

But

$$\bar{f}(a)(rx) = (\bar{f}(a) \circ r)(x) , \qquad \forall\, x \in R ,$$

consequently,

$$\bar{f}(ar) = \bar{f}(a) \circ r \qquad \forall\, r \in R .$$

Since $\bar{f} \in \mathrm{Hom}_R(A,F)$ extends f, we conclude that F_R is injective.

12. **THEOREM** (R. Baer). If M_R is any unital module, then there exists an injective unital module F_R and a monomorphism $\varphi\colon M_R \to F_R$.

PROOF. If $m \in M$, the correspondence $x \to mx$ is an element $m_0 \in H = \mathrm{Hom}_Z(R,M)$. Since $m_0 = 0$ if and only if $m = 0$, we have that $m \to m_0$ is an isomorphism of the additive groups $(M,+)$ and $(M_0,+)$, where $M_0 = \{m_0 \in H \mid m \in M\}$. Furthermore, H is an R-module (see the proof of 11), and $(mr)_0 = m_0 r \quad \forall\, m \in M, r \in R$, so that $m \to m_0$ is a monomorphism of M_R in H_R. Now if G is any abelian divisible group containing $(M,+)$, then

$$H_R = \mathrm{Hom}_Z(R,M) \subseteq F_R = \mathrm{Hom}_Z(R,G) .$$

Since F_R is injective by 11, $\varphi\colon M_R \to H_R$ is the desired monomorphism of M_R in F_R.

If M_R is an arbitrary module, whether or not R has an identity element, let R_1 denote the ring extension of R consisting of the set $\{(a,n) \mid a \in R, n \in Z\}$, where Z denotes the ring of integers, with addition

$$(a,n) + (b,m) = (a + b, n + m)$$

and multiplication

$$(a,n)(b,m) = (ab + nb + ma, nm) .$$

Then, as is well-known, R_1 is a ring with identity $(0,1)$ containing R as a subring, if we identify R with the subring $\{(a,0) \mid a \in R\}$ of R_1.

For each $x \in M_R$, and each $(a,n) \in R_1$, let

$$x(a,n) = xa + nx .$$

One easily verifies that M is an R_1-module under this operation, and we let M_{R_1} denote this module. Since $x(0,1) = x$, for all $x \in M$, it follows that M_{R_1} is a unital module.

Now if A is any R-module, trivially

$$\mathrm{Hom}_R(A,M) = \mathrm{Hom}_{R_1}(A,M) ,$$

and it follows from this that M_R is injective if and only if M_{R_1} is injective, or equivalently (1.2), if and only if M_{R_1} is u-injective. Now by 7, if M_R is any module, M_{R_1}, being unital, can be embedded in an injective module N_{R_1}. Then N_R is injective, and so:

13\. THEOREM. If M_R is any module, there exists an injective module N_R and a monomorphism $M_R \to N_R$.

Call a module M_R "set-injective" in case each monomorphism $0 \to M_R \to N_R$ splits. It is an easy matter to show that each injective module M_R is set-injective, since if M_R is injective the diagram below

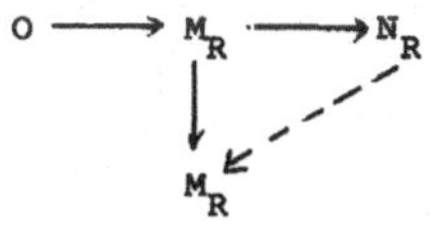

can be made commutative by a map indicated by ----→, and so by a remark in §0, $0 \to M_R \to N_R$ splits.

Conversely, suppose that M_R is "set-injective", and let $\varphi: M_R \to N_R$ be an embedding of M_R in an injective module given by 13. Then, $\varphi(M)$ is a direct summand of N_R, so $\varphi(M_R)$, hence M_R, is injective by 1.4. This completes the proof of the theorem below.

14. **THEOREM.** A module M_R is "set-injective" if and only if it is injective.

If M_R is any module, then we have noted that M_R is injective if and only if M_{R_1} is injective. But M_{R_1} is a unital module, so that M_{R_1} is injective if and only if M_{R_1} satisfies Baer's condition. This yields:

15. **PROPOSITION.** A module M_R is injective if and only if M_{R_1} satisfies Baer's condition.

Now assume that M_R is injective. Then, since any right ideal I of R is a right ideal of R_1, and since M_R satisfies Baer's condition we have

16. **COROLLARY.** If M_R is injective, then M_R satisfies Baer's condition. It is not hard to see that the converse of 16 fails, but we postpone the proof until the next section.

2. ESSENTIAL EXTENSIONS AND THE INJECTIVE HULL

We recall that a module B_R is an *extension* of a module A_R in case there exists a monomorphism $f: A_R \to B_R$; we denote this extension by (B,A,f) or by $B \geq_f A$. The set of all extensions of a module A_R is non-empty since $A \geq_i A$, where i denotes the injection map $A \to A$.

If (B,A,f) and (C,A,g) are two extensions of A_R , then we write $(B,A,f) \geq_\varphi (C,A,g)$, or $(C,A,g) \leq_\varphi (B,A,f)$, in case there is a monomorphism $\varphi: C \to B$ such that $f = \varphi g$, that is, such that the diagram

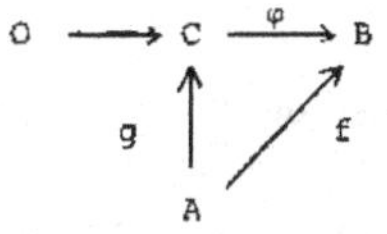

is commutative. One easily verifies that the relation $\geq$ defined between certain extensions of a given module A_R is reflexive and transitive. If $(B,A,f) \geq_\varphi (C,A,g)$ and if $\varphi: C \to B$ is an epimorphism then we write $(B,A,f) =_\varphi (C,A,g)$. It follows that "=" is an equivalence relation. An extension $B \geq_f A$ of a module A_R is proper, and we write $B >_f A$, in case $(B,A,f) \neq_f (A,A,i)$, where $i: A \to A$ is the injection map. We denote (A,A,i) by simply A .

We reserve the symbol $B \geq A$ to denote extensions as defined above. In case A is a submodule of B , we write $B \supseteq A$; $B \supset A$ signifies that $B \supseteq A$ and $B \neq A$.

If M_R is a submodule of a module B_R , then we say that M_R is an *essential submodule*, and write $B_R' \supseteq M_R$, in case each nonzero submodule C of B meets M_R , that is, $C \cap M \neq 0$. It is the same thing to require that the submodule generated by any nonzero element of B meets M .

If $B \geq_f A$, then we say that B *is an essential extension of* A , and write $B' \geq_f A$, in case $f(A)$ is an essential submodule of B . Each module A_R has at least one essential extension, namely itself. On the other hand, an injective module M_R can have no proper essential extension. For if $N' \geq_f M$, then, since M_R is injective, there exists a submodule B of N such that $N = f(M) \oplus B$, and then, since

$f(M) \leq' N$, we must have $B = O$, that is, $N = f(M)$. Thus, $(N,M,f) = {}_f(M,M,i)$, and so (N,M,f) is not a proper extension.

Occasionally, we write $O \longrightarrow A \xrightarrow{f}{}' B$ to denote $B' \geq_f A$.

If $B_R \supseteq A_R$, then K is a <u>complement submodule</u> of A in B in case K is a submodule which is maximal in the set of all submodules Q such that $Q \cap A = 0$. By Zorn's lemma, if P is any submodule of B satisfying $A \cap P = O$, then there exists a complement submodule of A in B which contains P . Any submodule K of B which is a complement in B of some submodule of B is called a <u>complement submodule of</u> B . Note that O and B are complement submodules of B , and each is the unique complement of the other in B .

1. REMARKS

1. (a) If P_R, N_R, and M_R are modules, and if $P \leq (N,P,g) \leq_\theta (M,P,f)$ then $P \leq' (M,P,f)$ if and only if $N \leq' (M,N,\theta)$ and $P \leq' (N,P,g)$;

(b) If $P \leq' (N,P,g)$ and if $(N,P,g) = (M,P,f)$ then $P \leq' (M,P,f)$.

PROOF. First of all the following diagram is commutative

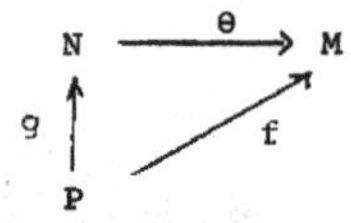

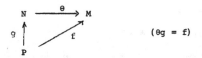

$(\theta g = f)$

where g, f, and θ are monomorphisms. We assume now that $P \xrightarrow{f}{}' M$, and prove that this implies $N \xrightarrow{\theta}{}' M$ and $P \xrightarrow{g}{}' N$: If K is any nonzero submodule of M , then $K \cap f(P) \neq O$. But $\theta g = f$, so $K \cap \theta[g(P)] \neq O$ so that $K \cap \theta(N) \neq O$; and hence $N \longrightarrow{}' M$. Next, if Q is any nonzero submodule of N , then $\theta(Q) \neq O$, so $P \xrightarrow{f}{}' M$ implies $\theta(Q) \cap f(P) \neq O$ and $Q \cap \theta^{-1}f(P) \neq O$. Since $g = \theta^{-1}f$ on P , this yields $Q \cap g(P) \neq O$, so that $P \xrightarrow{g}{}' N$.

Conversely assume $N \xrightarrow{\theta}{}' M$ and $P \xrightarrow{g}{}' N$, and let K be a nonzero submodule of M . Then $T = K \cap \theta(N) \neq O$, so that $\theta^{-1}T \neq O$, and then $\theta^{-1}T \cap g(P) \neq O$. Thus, $T \cap \theta g(P) \neq O$ and $K \cap \theta g(P) \neq O$. Since $\theta g(P) = f(P)$, we have $K \cap f(P) \neq O$, so that $P \xrightarrow{f}{}' M$.

The last assertion now follows, since if $(N,P,g) = {}_{\theta}(M,P,f)$, then trivially $N \leq' (M,N,\theta)$, and we can apply (a) to conclude that $P \leq' (M,P,f)$.

2. Let P,N be submodules of M. Then $M' \supseteq (P \cap N)$ if and only if $M' \supseteq N$ and $M' \supseteq P$.

PROOF. If $M \,'\!\supseteq (P \cap N)$, then $M \,'\!\supseteq N$ and $M \,'\!\supseteq P$ by 1. Conversely, if $M \,'\!\supseteq N$ and $M \,'\!\supseteq P$, then any nonzero submodule K of M satisfies $K \cap N \neq 0$, and then $K \cap (N \cap P) = (K \cap N) \cap P \neq 0$. This shows that $M \,'\!\supseteq (N \cap P)$.

3. Let N be a submodule of M. Then there exists a submodule Q, $N \subseteq Q \subseteq M$, such that Q is a maximal essential extension of N contained in M.

PROOF. We give two proofs. The first is the more obvious one: a maximal essential extension exists by Zorn's lemma. Another proof follows from 3' below.

3'. If N is a submodule of M, and if K is any complement of N in M, then there exists a complement Q of K in M such that $Q \supseteq N$. Furthermore, any such Q is a maximal essential extension of N in M.

PROOF. Since $N \cap K = 0$, by Zorn's lemma there exists a complement Q of K in M which contains N. Let T be any nonzero submodule of Q and suppose for the moment that $T \cap N = 0$, and consider the submodule $K_1 = T + K$. Since $T \cap K = 0$, $K_1 > K$. Suppose $x = t + k \in (T + K) \cap N$, where $t \in T$, and $k \in K$. Then $k = x - t \in Q \cap K$, so $k = 0$, and $x = t \in N \cap T$. Since $N \cap T = 0$, we see that $x = 0$, so that $(T + K) \cap N = 0$. Since $K_1 = T + K > K$, this violates the choice of K as a complement of N. Hence $T \cap N \neq 0$, and $Q \,'\!\supseteq N$. Now let P be a submodule of M properly containing Q. Then $P \cap K \neq 0$ and $(P \cap K) \cap N = P \cap (K \cap N) = P \cap 0 = 0$. Thus, P is not an essential extension of N. This shows that Q is a maximal essential extension of N.

If M_R is any module, a submodule N_R is <u>closed</u> in case N_R has no proper essential extensions in M_R. An immediate consequence of 3' is that any closed submodule N of M is a complement submodule (the complement submodule Q must coincide with N, if N is closed). Furthermore, 3' states that if N is a submodule of M, and if K is a complement of N in M, then <u>any</u> complement of K in M which contains N is a maximal essential extension of N in M. Now assume that N itself is a complement submodule, say N is the complement of a submodule P of M. Then, there exists a complement K of N which contains P, and clearly N is also a complement of K. Then, by 3' we conclude that N is closed. We have proved:

4. The closed submodules of a module M coincide with the complement submodules of M. Furthermore, if N and K are complement submodules, and if K is a complement of N in M, then N is a complement of K in M.

5. Let N be a submodule of M, let K be any complement of N in M, and let $N_1 = N + K$. Then $M\ '\supseteq N_1$ and $(M - K)\ '\supseteq (N_1 - K)$.

PROOF. Let Q be any nonzero submodule of M. If $Q \subseteq K$, then $Q \cap N_1 = Q \neq 0$. If $Q \not\subseteq K$, and then $Q + K \supset K$, so $(Q + K) \cap N \neq 0$. If $0 \neq x = q + k \in (Q + K) \cap N$, $q \in Q$, $k \in K$, then $K \cap N = 0$ implies $q \neq 0$, and then $0 \neq q = x - k \in N_1 \cap Q$. Thus $M\ '\supseteq N_1$. In particular, if $Q \supset K$, we have that $N_1 \cap Q \supset K$. This shows that $(Q - K) \cap (N_1 - K) \neq 0$, and so $(M - K)\ '\supseteq N_1 - K$.

Call a module M_R essentially-injective in case, for each module A, each row-exact diagram on the left can be embedded in a commutative diagram on the right.

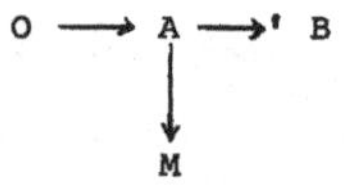

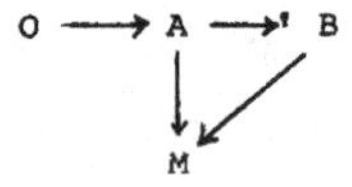

6. Any essentially-injective module is injective.

PROOF. Let C_R be any module and let A_R be any submodule. Furthermore, let K denote any complement of A in C. Since the sum $B = A + K$ is direct, if $f: A \to M$ is any map of A into M, f can be extended to a map $g: B \to M$. Since $C\ '\supseteq B$ by 5, and since M_R is essentially-injective, g can be extended to a map $h: C \to M$. Since h extends f, this shows that M_R is injective.

7. If M_R is injective, then any row- and column-exact diagram on the left

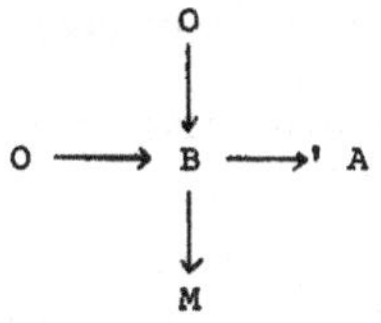

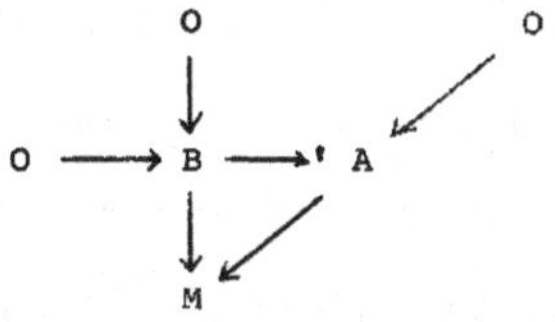

can be embedded in a commutative diagram on the right, where $O \to A \to M$ is exact. Expressed otherwise: <u>any monomorphism of a module</u> B <u>in an injective module</u> M <u>can be extended to a monomorphism into</u> M <u>of any essential extension of</u> B .

PROOF. Since M_R is injective, the diagram on the left above can be embedded in a commutative diagram below.

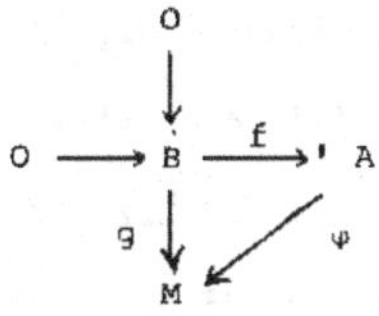

If $x \in Ker(\varphi) \cap f(B)$, then $g(f^{-1}(x)) = \varphi(x) = O$, so that $f^{-1}(x) = O$, and so $f(f^{-1}(x)) = x = O$. Thus, $Ker(\varphi) \cap f(B) = O$. Since $A\ '\geq_f B$, this implies that $Ker(\varphi) = O$, and the proof is complete.

Let $\xi^*(M_R)$ denote the totality of all essential extensions of M_R . An extension $(A,M,f) \in \xi^*(M_R)$ is said to be a <u>maximal essential extension</u> in case (1) below holds:

(1) If $(B,M,g) \in \xi^*(M)$, then $(B,M,g) \leq_\varphi (A,M,f)$ for some monomorphism $\varphi: B \to A$.

Let $\xi(M_R)$ denote the (possibly empty) collection of all maximal essential extensions of M_R . Below we establish the following proposition:

(1') If $(A,M,f) \in \xi(M)$, if $(C,M,h) \in \xi^*(M)$, and if $(C,M,h) \geq_\theta (A,M,f)$, then $(C,M,h) =_\theta (A,M,f)$.

An extension (A,M,f) of M is injective, in case A is injective. Let $\vartheta^*(M)$ denote the totality of all injective extensions of M . Then $(A,M,f) \in \vartheta^*(M)$ is a <u>minimal injective extension</u> in case (2) below is satisfied:

(2) If $(B,M,g) \in \vartheta^*(M)$, then $(B,M,g) \geq_{\varphi} (A,M,f)$ for some monomorphism $\varphi: A \to B$.

Let $\vartheta(M)$ denote the (possibly empty) collection of minimal injective extensions of M. Below we establish the following proposition:

(2') If $(A,M,f) \in \vartheta(M)$, if $(C,M,h) \in \vartheta^*(M)$, and if $(C,M,h) \leq_{\theta} (A,M,f)$, then $(C,M,h) =_{\theta} (A,M,f)$.

The main aim of this section is to show that any module has a minimal injective extension, and that any two such are equivalent. The following lemma somewhat clarifies the situation.

8. LEMMA. Let M_R be such that $\xi^*(M) \cap \vartheta^*(M)$ is non-empty. Then: (i) $\xi(M) = \vartheta(M) = \xi^*(M) \cap \vartheta^*(M)$; (ii) $\xi(M) = \vartheta(M)$ consists of equivalent extensions; (iii) each $(A,M,f) \in \xi(M) = \vartheta(M)$ satisfies (1') and (2').

PROOF. Let $(A,M,f) \in \xi^*(M) \cap \vartheta^*(M)$, and let $(B,M,g) \in \xi^*(M)$. Then, since A is injective, by 7 there exists a monomorphism $\varphi: B \to A$ such that $(B,M,g) \leq_{\varphi} (A,M,f)$. Thus (A,M,f) satisfies (1), and so $\xi(M) \supseteq \xi^*(M) \cap \vartheta^*(M)$. If, on the other hand, $(B,M,g) \in \vartheta^*(M)$, then since $(A,M,f) \in \xi^*(M)$ and B is injective, by 7 again there exists a monomorphism $\varphi: A \to B$ such that $(A,M,f) \leq_{\varphi} (B,M,g)$. Thus, (A,M,f) satisfies (2), and so $\vartheta(M) \supseteq \xi^*(M) \cap \vartheta^*(M)$. Then $\xi(M) \cap \vartheta(M) \supseteq \xi^*(M) \cap \vartheta^*(M)$. Since $\xi(M) \subseteq \xi^*(M)$, and $\vartheta(M) \subseteq \vartheta^*(M)$, we conclude that $\xi(M) \cap \vartheta(M) = \xi^*(M) \cap \vartheta^*(M)$.

Next we show that $(A,M,f) \in \xi(M) \cap \vartheta(M)$ satisfies (1'). If $(C,M,h) \in \xi^*(M)$ is given as in (1'), it follows from 1 that $(C,M,h) \,'\!\geq_{\theta} (A,M,f)$. Since A is injective, $\theta(A)$ is a direct summand of C, so that necessarily $\theta(A) = C$, and $(C,M,h) =_{\theta} (A,M,f)$. Thus, (A,M,f) satisfies (1'). In particular, if $(C,M,H) \in \xi(M)$, then (1) asserts the existence of a monomorphism $\varphi: C \to A$ such that $(C,M,h) \leq_{\varphi} (A,M,f)$. Then, applying (1'), we conclude that $(C,M,h) =_{\varphi} (A,M,f)$. This implies that $(C,M,h) \in \xi(M) \cap \vartheta(M)$, and so $\xi(M) \subseteq \vartheta(M)$. Furthermore this also shows that $\xi(M)$ consists of equivalent extensions (each equivalent to (A,M,f)).

Finally we show that $(A,M,f) \in \xi(M) \cap \vartheta(M)$ satisfies (2'). If $(C,M,h) \in \vartheta^*(M)$ is given as in (2'), then since $A' \geq_f M$, then $(A,M,f) \,'\!\geq_{\theta} (C,M,h)$ by 1. But $\theta(C)$ is injective, hence a direct summand of A, so we conclude that $\theta(C) = A$ and that

$(A,M,f) =_{\theta} (C,M,h)$. This shows that (A,M,f) satisfies (2'). In particular, if $(C,M,h) \in \vartheta(M)$, then (2) asserts the existence of a monomorphism φ such that $(C,M,h) \leq_{\varphi} (A,M,f)$. Then applying (2') we conclude that $(C,M,h) =_{\varphi} (A,M,f)$, showing that $\vartheta(M) \subseteq \xi(M) \cap \vartheta(M)$, and that $\vartheta(M) = \xi(M)$. This concludes the proof of (i), (ii), and (iii).

9. THEOREM. If M_R is any module, the collection $\xi(M_R)$ of maximal essential extensions of M_R is non-empty, and coincides with the collection of minimal injective extensions of M_R. (cf. 8).

PROOF. By 8, it suffices to show that $\xi(M) \cap \vartheta(M)$ is non-empty. By §1, M has an injective extension (P,M,h), and by 4, there exists an essential extension N of $h(M)$ in P such that N is a closed submodule of P. For example, if K is any complement submodule of $h(M)$ in P, and if N is a complement of K in P which contains $h(M)$, then by 3' and 4, N is closed and $N \;'\!\supseteq h(M)$. Now K is a complement of N in P, and by 6, $P - K \;'\!\supseteq N_1 - K$, where $N_1 = N + K$. If η denotes the natural isomorphism $N \cong N_1 - K$, and if $\varphi = \eta^{-1}$, then by 7, the diagram below

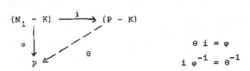

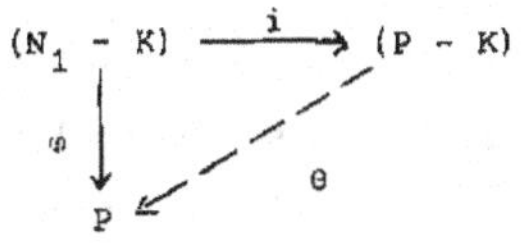

$$\theta i = \varphi$$
$$i\varphi^{-1} = \theta^{-1}$$

can be completed by a monomorphism θ indicated by $--\rightarrow$. By (a) of 1, we see that $P - K \;'\!\supseteq N_1 - K$ implies $\theta(P - K) \;'\!\supseteq \theta(N_1 - K)$, i.e., $\theta(P - K) \;'\!\supseteq N$. Since N is closed in P, we conclude that $\theta(P - K) = N$. Then

$$P - K = \theta^{-1}N = i\,\varphi^{-1}(N) = \theta^{-1}(N) = N_1 - K .$$

From this we conclude that $P = N + K = N \oplus K$. Injectivity of N now follows from that of P. Thus, using notation of 8, we have that $(N,M,h) \in \xi^*(M) \cap \vartheta^*(M)$, and that $\xi(M) = \vartheta(M) = \xi^*(M) \cap \vartheta^*(M)$.

We let $E(M_R)$ denote any of the mutually equivalent extensions in $\xi(M_R) = \vartheta(M_R)$, and call it the <u>injective hull of</u> M_R.

10. PROPOSITION. (1) If (A,M,f) is an essential extension of M, then $\xi(A) = \xi(M)$; (2) If φ is any isomorphism of M_R, then $\xi(M) = \xi(\varphi(M))$.

PROOF. (1) Each injective extension (Q,A,g) of A is an injective extension (Q,M,gf) of M. By 7, the converse is true: each injective extension (P,M,h) of M is an injective extension (P,A,φ) of A, for some monomorphism $\varphi: A \to P$ such that $(P,M,h) \supseteq_{\varphi} (P,M,g)$. From these facts follows that the minimal injective extensions of each coincide. (2) follows (1) since $(\varphi(M),M,\varphi) \;'\!\supseteq M$.

11. PROPOSITION. Let M_R be injective, and let P and N be essential submodules of M_R. Then $P \cong N$ if and only if there exists an automorphism of M which maps P onto N.

PROOF. The "if" part is trivial. Now assume that $P \cong_{\varphi} N$, and let φ' denote an extension of φ to $\mathrm{Hom}_R(M,M)$. Since $\mathrm{Ker}(\varphi') \cap P = 0$, φ' is a monomorphism. Since $\varphi'(M)$ is therefore injective, since $M \supseteq_{\varphi'}(M) \supseteq \varphi(P) = N$, and since $M \;'\!\supseteq N$, it follows that $M = \varphi'(M)$, and, hence, φ' is an automorphism of M which induces φ.

We now show that Baer's criterion (see §1) fails for arbitrary modules. Recall that an R-module P is trivial in case $PR = 0$. If P is any trivial R-module, let $\bar{P}$ denote the trivial R-module, where $\bar{P}$ is the injective hull of the additive group $(P,+)$, that is, $(\bar{P},+)$ is the smallest divisible group containing $(P,+)$. If $0 \neq x \in \bar{P}$, then the submodule generated by x is Zx, since $xR = 0$. Since $\bar{P}_Z$ is the injective hull of P_Z, then $\bar{P}_Z \;'\!\supseteq P_Z$, and necessarily $Zx \cap P \neq 0$. Thus $\bar{P}_R$ is an essential extension of P_R.

Now let R by any ring with identity 1, let N_R be any u-injective module, and let P_R be any trivial module such that $(P,+)$ is not divisible. If $M_R = N_R \oplus P_R$, and if f is any module homomorphism of a right ideal I of R into M_R, then $f(I) = f(I)1 \subseteq N_R$. Then by Baer's criterion for unital modules, there exists $m \in N$ such that $f(x) = mx \;\forall\, x \in I$. Thus, M_R satisfies Baer's condition. However, M_R is not injective, since if $\bar{P}_R$ denotes the trivial R-module defined above, then $Q_R = N_R \oplus \bar{P}_R$ is an essential extension of M_R properly containing M_R, since $\bar{P}_R$ is a proper essential extension of P_R. Hence Baer's criterion fails.

EXERCISE

Two injective modules which are isomorphic to submodules of each other are isomorphic (R. Bumby and B. Osofsky). If you get stuck see Bumby's Abstract 601-15, Notices of the Amer. Math. Soc. 10 (1963) p. 273. The proof is patterned after the corresponding result of set theory, the Cantor-Bernstein theorem. (See Bumby's note, Modules which are isomorphic to submodules of each other, Arch. Math. 16 (1965) 184-185).

3. QUASI-INJECTIVE MODULES

A module M_R is said to be <u>quasi-injective</u> in case each homomorphism of any submodule N into M can be extended to a homomorphism of M into M. A module M_R is simple in case 0 and M are the only submodules of M. Injective modules and simple modules are then examples of quasi-injective modules. However, there exist quasi-injective modules which are neither injective nor simple.

EXAMPLE. Let Z_{p^n} denote the cyclic group of order p^n, where p is a prime, and let N be a divisible group containing Z_p (cf. §1). If x_o denotes a generator of Z_p, let x_i be elements *) of N defined inductively as follows:

$$px_1 = x_o \, , \; px_2 = x_1, \; \ldots, \; px_i = x_{i-1}, \; \ldots \quad (i = 1,2,\ldots).$$

Clearly x_i generates a subgroup of N isomorphic to $Z_{p^{i+1}}$, $\forall_i$, and $Z_{p^o} \subset Z_p \subset \ldots \subset Z_{p^i} \subset \ldots$. Let P denote the set theoretical union of these groups. If $(n,p) = 1$, then $nZ_{p^i} = Z_{p^i}$, so $nP = P$. Furthermore, $p^k Z_{p^{i+k}} = Z_{p^i}$ $\forall$ i, k, so that $p^k P = P$ $\forall$ k. This shows that P is a divisible group. If $y \in Z_{p^{k+1}}$, $y \notin Z_{p^k}$, then $0 \neq py \in Z_{p^k}$, so that $Z_{p^{k+1}}$ is an essential extension of Z_{p^k}, $k = 1,2,\ldots$. It follows that P is an essential extension of Z_{p^k} $\forall$ k, so that P_Z, being injective, is the injective hull of Z_{p^k} $\forall$ k. For brevity, denote Z_{p^k} by P_k. It is easy to see that P_k consists of all elements of P of order $\leq p^k$, so that $\lambda P_k \subseteq P_k$ $\forall \lambda \in \Lambda$, where $\Lambda = \mathrm{Hom}_Z(P,P)$. Now if S is any submodule of P_k, and if $f: S \to P_k$ is any homomorphism of S into P_k, then by injectivity of P_Z, f is induced by an element $\lambda \in \Lambda$. However λ induces an element $\bar{\lambda}$ of $\mathrm{Hom}_Z(P_k,P_k)$ which induces f. Hence P_k is quasi-injective $\forall$ k. However, if $k > 1$, then P_k contains the subgroup $P_{k-1} \neq P_k$, 0, so these quasi-injective modules are not simple. Furthermore, $E(P_k) = P \neq P_k$, so these modules are not injective.

Quasi-injective and injective modules have the following simple relationship, first noted by Johnson and Wong [1].

---

*) Since N is divisible, the elements x_i exist but are not in general unique.

1. PROPOSITION. If M_R is any module, if $\hat{M} = E(M)$, and if $\Lambda = \mathrm{Hom}_R(\hat{M},\hat{M})$, then: (1) $\Lambda\hat{M}$ is the intersection of all quasi-injective submodules of $\hat{M}$ containing M; (2) $\Lambda\hat{M}$ is quasi-injective; (3) M is quasi-injective if and only if $M = \Lambda\hat{M}$.

PROOF. (2) If $f: N \to \Lambda\hat{M}$ is any map of a submodule N of $\Lambda\hat{M}$ into $\Lambda\hat{M}$, then f is induced by some $\lambda \in \Lambda$. Since $\lambda(\Lambda\hat{M}) \subseteq \Lambda\hat{M}$, λ induces $\bar{\lambda} \in \mathrm{Hom}_R(\Lambda\hat{M},\Lambda\hat{M})$, and $\bar{\lambda}$ also induces f, showing that $\Lambda\hat{M}$ is quasi-injective.

(1) Let P be any quasi-injective submodule of $\hat{M}$ containing M. We wish to show that $P \supseteq \Lambda\hat{M}$, so it is sufficient to show that $\alpha P \subseteq P \ \forall\, \alpha \in \Lambda$. To do this, we note that $Q(\alpha) = \{x \in P \mid \alpha x \in P\}$ is a submodule of P, and we have only to show that $Q(\alpha) = P \ \forall\, \alpha \in \Lambda$. Since $q \to \alpha q$, $q \in Q = Q(\alpha)$ a map of Q into P, and since P is quasi-injective, there exists $\alpha_1 \in \mathrm{Hom}_R(P,P)$ such that $\alpha_1 q = \alpha q \ \forall\, q \in Q$. Since $\hat{M}$ is injective, there exists $\alpha' \in \Lambda$ such that $\alpha' x = \alpha_1 x \ \forall\, x \in P$. Since $\alpha' P \subseteq P$, if $(\alpha' - \alpha)P = 0$, we have $\alpha P \subseteq P$. Thus, if $Q(\alpha) \neq P$, then $(\alpha' - \alpha)P \neq 0$. Since $M\ '\supseteq M$, necessarily $\hat{M}\ '\supseteq P$, and consequently $(\alpha' - \alpha)P \cap P \neq 0$. But if x, $0 \neq y \in P$ are such that $y = (\alpha' - \alpha)x \in (\alpha' - \alpha)P \cap P$, then since $\alpha' x = \alpha_1 x \in P$, we have that $\alpha x = \alpha_1 x - y \in P$. But then $x \in Q(\alpha)$, so that $\alpha x = \alpha_1 x$, and so $y = 0$, a contradiction which establishes (1). (3) is an immediate consequence of (1) and (2).

We remark without proof that if $M_R \cong N_R$, then M_R is quasi-injective if and only if N_R is.

A submodule N of M is closed in case each submodule of M which contains N and is essential over N coincides with N (see §2).

2. PROPOSITION. Let M_R be quasi-injective, and let N be a closed submodule. Then, any map w of a submodule K of M into N can be extended to a map u of M into N.

PROOF. By Zorn's lemma we can assume that K is such that w cannot be extended to a map of T into N for any submodule T of M which properly contains K. Since M_R is quasi-injective, w is induced by a map $u: M \to M$. Suppose $u(M) \not\subseteq N$, and let L be a complement of N in M. Since N is closed, N is a complement of L. Therefore, since $u(M) + N \supset N$, we see that $(u(M) + N) \cap L \neq 0$. Let $0 \neq x = a+b \in (u(M)+N) \cap L$, $a \in u(M)$, $b \in N$. If $a \in N$, then $x \in N \cap L = 0$, a contradiction. Therefore, $a \notin N$, and $a = x-b \in L \oplus N$. Now $T = \{y \in M \mid u(y) \in L \oplus N\}$ is a submodule of M containing K. If $y \in M$ is such that $u(y) = a$, then $y \in T$, but $y \notin K$, since $a \notin N$. Let π denote the projection of $L \oplus N$ on N. Then πu is a map of T in N, and $\pi u(y) = u(y) = w(y)\ \forall\ y \in K$. Thus πu is a proper extension of w, a contradiction. Therefore, $u(M) \subseteq N$, and u is the desired extension of w.

Analogously to the corresponding result for injective modules, one proves that if a direct product $\Pi_{i\in I}M_i$ of R-modules $\{M_i \mid i \in I\}$ is quasi-injective then M_i is quasi-injective $\forall\ i \in I$. Let $M = Q \oplus Z_p$, where p is a prime, and Q is the additive group of rational numbers. Then the canonical epimorphism $\lambda: Z \to Z_p$ is a map of the subgroup Z of Q onto Z_p which can not be extended to $\mathrm{Hom}_Z(Q,Z_p)$, and therefore cannot be extended to $\mathrm{Hom}_Z(M,M)$. Thus a sum of quasi-injective modules need not be quasi-injective.

An extension (P,M,f) of a module M_R is a <u>minimal quasi-injective extension</u> in case P_R is quasi-injective and in case the following condition is satisfied:

(1) If (A,M,g) is any quasi-injective extension of M, then there exists a monomorphism $\varphi: P \to A$ such that $(P,M,f) \leq_{\varphi} (A,M,g)$.

3. COROLLARY. Let M_R be quasi-injective. (1) If N is any closed submodule of M, then N is a direct summand of M, and N is quasi-injective. (2) If P is any submodule of M, then there exists a quasi-injective essential extension of P contained in M. (3) Each minimal quasi-injective extension of a module K_R is an essential extension of K_R.

PROOF. (1) If $e: M \to N$ is the extension given by the theorem of the injection map $N \to N$, then $M = N \oplus \mathrm{Ker}(e)$, so that N is a direct summand of M; N is quasi-injective by the remark preceding the corollary.

(2) By Zorn's lemma, P is contained in a closed submodule N which is an essential extension of P, and N is quasi-injective by (1). (3) is an immediate consequence of (2).

4. THEOREM. In the notation of 1, $(\Lambda M,M,i)$ is a minimal quasi-injective extension of M, where i denotes the injection $M \to \Lambda M$. Any two minimal quasi-injective extensions are equivalent.

PROOF. Let (A,M,g) be any quasi-injective extension of M, let $\hat{A} = E(A)$, and let $\Omega = \mathrm{Hom}_R(\hat{A},\hat{A})$. Then, by 1, $\Omega A \subseteq A$. Since $M_o = \Lambda M$ is an essential extension of M, the monomorphism $g: M \to \hat{A}$ can be extended to a monomorphism (also denoted by g) of M_o in A. Since $g(M_o)$ is quasi-injective, then $\Omega(g(M_o)) \subseteq g(M_o)$, and we conclude that $\Omega(B) \subseteq B$, where $B = A \cap g(M_o)$. Then, by 1, B is quasi-injective. It follows that $g^{-1}B$ is a quasi-injective extension of $M \subseteq M_o = \Lambda M$. Since ΛM is the smallest quasi-injective extension of M contained in $\hat{M}$, we conclude that $g^{-1}B = M_o$, so $B = g(M_o) \subseteq A$. This establishes that $M_o = \Lambda M$ is a minimal quasi-injective extension. It follows immediately that if (A,M,g) is also a minimal quasi-injective extension of M, that (A,M,g) is equivalent to ΛM.

In the future $Q(M)$ will denote any of the equivalent minimal quasi-injective extensions of M. By 3 and 4, we have:

5. COROLLARY. Let M be a quasi-injective module and let N be a submodule. Then $M = Q(N)$ if and only if $M \,'\!\supseteq N$.

6. LEMMA. Let M be a module, let $\{M_i \mid i \in I\}$ be a family of independent submodules, and let Q_i be an essential extension of M_i in $M \,\forall\, i \in I$. Then $\{Q_i \mid i \in I\}$ is an independent family of submodules, and $\Sigma_{i\in I} Q_i$ is an essential extension of $\Sigma_{i\in I} M_i$.

PROOF. Let $K_i = \Sigma_{i\neq j\in I} M_j$. Since $M_i \cap K_i = 0$, $(Q_i \cap K_i) \cap M_i = 0 \;\forall\; i \in I$. Since $Q_i \,'\!\supseteq M_i$, this shows that $Q_i \cap K_i = 0 \;\forall\; i \in I$, so that the augmented family $\{Q_i, M_j \mid i \neq j \in I\}$ for a fixed $i \in I$ is also independent. Let P denote the set of all subsets J of I such that the family $F_J = \{Q_j, M_i \mid j \in J,\, i \in I-J\}$ is independent, and let $\{J_\lambda \mid \lambda \in \Lambda\}$ be a chain in P (ordered by inclusion). If $J = \cup_{\lambda\in\Lambda} J_\lambda$, it is easy to see that F_J is independent, since each finite subfamily of F_J is independent. By Zorn's lemma, P has a maximal element, call it J. Then the proof above for the case of the fixed i shows that $J = I$. Thus $\{Q_i \mid i \in I\}$ is independent.

Let π denote the projection of $Q_i \oplus K_i$ on Q_i, where $K_i = \Sigma_{i\neq j\in I} M_j$ as before. If P is a nonzero submodule of $Q_i \oplus K_i$, then $\pi(P) = 0$ implies that $P \subseteq K_i$, so $P \cap (M_i \oplus K_i) \neq 0$. If $\pi(P) \neq 0$, then $\pi(P) \cap M_i \neq 0$, and then $P \cap M_i \oplus K_i \neq 0$.

This shows that $Q_i + K_i \ '\supseteq \Sigma_{j \in I} M_j$ for an arbitrary $i \in I$. By applying Zorn's lemma similarly as above, we conclude that $\Sigma_{i \in I} Q_i$ is an essential extension of $\Sigma_{i \in I} M_i$.

7. PROPOSITION. Let M be any injective module, let $\{M_i \mid i = 1, \dots, n\}$ be any finite family of independent submodules of M , and for each i choose in M an injective hull $E_i = E(M_i)$ of M_i . Then E_i is a family of independent submodules, and

$$E(\Sigma_1^n \oplus M_i) = \Sigma_1^n \oplus E(M_i)$$

PROOF. By §2, we know we can choose $E(M_i)$ to be any maximal essential extension of M_i contained in M. By 6, the sum $\Sigma_1^n E_i$ is direct, and is an essential extension of $\Sigma_1^n \oplus M_i$. Since a direct sum of finitely many injective modules is injective, $\Sigma_1^n \oplus E_i = E(\Sigma_1^n \oplus M_i)$.

REMARK. If M is assumed only to be quasi-injective, by 3, we can choose $Q_i = Q(M_i) \subseteq M$, $i = 1, \dots, n$. Then by 3, (also 4) $Q_i \ '\supseteq M_i$, so by 6, the sum $Q = \Sigma_1^n Q_i$ is direct, and $Q \ '\supseteq \Sigma_1^n \oplus M_i$. It follows that $Q(\Sigma_1^n \oplus M_i) \supseteq \Sigma_1^n \oplus Q(M_i)$, and the example preceding 3, shows that the inclusion may be proper.

If M_R is a module, for any subset $N \subseteq M$ let $N^R = \{r \in R \mid Nr = 0\}$; for any subset $S \subseteq R$, let $S^M = \{m \in M \mid mS = 0\}$; and set $N^{RM} = (N^R)^M$, $S^{MR} = (S^M)^R$. The set N^R is always a right ideal of R, while S^M is always a fully invariant submodule of M (in the sense that $\Lambda S^M \subseteq S^M$, where $\Lambda = \mathrm{Hom}_R(M,M)$). Also: $N^{RM} \supseteq N$ and $S^{MR} \supseteq S$ $\forall N \subseteq M$, $S \subseteq R$.

8. THEOREM.*) Let M_R be a quasi-injective module such that $R^M = 0$, and let $\Lambda = \mathrm{Hom}_R(M,M)$. Then: (1) If N is any Λ-submodule of ${}_\Lambda M$ satisfying $N^{RM} = N$, then

$$(N + \Lambda x)^{RM} = N + \Lambda x \qquad \forall x \in M .$$

(2) If $x_1, \dots, x_n \in M$, then

---

*) Cf. Jacobson [2, p.27, Lemma] and Johnson-Wong [1].

$$(\Sigma_{i=1}^{n} \Lambda x_i)^{RM} = \Sigma_{i=1}^{n} \Lambda x_i .$$

PROOF. Let $A = N^R$, and $B = x^R$. By assumption $A^M = N^{RM} = N$. Clearly $B \subseteq (\Lambda x)^R$, but if $r \in (\Lambda x)^R$, then $\Lambda xr = 0$, and so $xr = 0$, implying that $r \in (x)^R$ and that $B = x^R = (\Lambda x)^R$.

(1) Since $P^{RM} \supseteq P$ for each subset $P \subseteq M$, we have that $(N + \Lambda x)^{RM} \supseteq N + \Lambda x$, and therefore it suffices to show that $(N + \Lambda x)^{RM} \subseteq N + \Lambda x$. Since $(N + \Lambda x)^R = N^R \cap (\Lambda x)^R = A \cap B$, we must show that $(A \cap B)^M \subseteq N + \Lambda x$. To do this, let $y \in (A \cap B)^M$, and consider the correspondence $\theta: xa \to ya$, $a \in A$, between the modules xA and yA. If $xa = xb$, $a, b \in A$, then $x(a-b) = 0$, and so $(a-b) \in A \cap B$. Since $y \in (A \cap B)^M$, necessarily $y(a-b) = 0$, and so $ya = yb$. Thus θ is a single-valued function. Clearly $\theta(xa)r = \theta(xar)$ $\forall r \in R$, so that $\theta \in \mathrm{Hom}_R(xA, yA)$. Since M_R is quasi-injective, there exists $\lambda \in \Lambda$ such that λ induces θ, that is, such that $\lambda xa = ya$ $\forall a \in \Lambda$. But then $(\lambda x - y)A = 0$, and so $\lambda x - y \in N = A^M$, which shows that $y \in N + \Lambda x$, proving that $(A \cap B)^M \subseteq N + \Lambda x$.

(2) If $x_1, \ldots, x_n \in M$, then taking $N = 0$, by (1) we have $(0 + \Lambda x_1)^{RM} = 0 + \Lambda x_1$. By induction we can assume that $P^{RM} = P$, where $P = \Sigma_{i=1}^{n-1} \Lambda x_i$. Then (1) establishes that $(P + \Lambda x_n)^{RM} = P + \Lambda x_n$, proving (2).

If M_R is any module, then $x_1, \ldots, x_n \in M$ are linearly independent over Λ in case $x_i \notin N_i$, $i = 1, \ldots, n$, where $N_i = \Sigma_{j \neq i} \Lambda x_j$, $i = 1, \ldots, n$.

9. PROPOSITION. Under the assumptions of 8, if $x_1, \ldots, x_n \in M$ are linearly independent over Λ, then: (1) There exist right ideals A_i, $i=1,\ldots,n$, of R such that $x_i A_i \neq 0$ and $x_i A_j = 0$, $j \neq i = 1,\ldots,n$. (2) If $y_i \in x_i A_i$, then there exists $a \in R$ such that $x_i a = y_i$, $i = 1,\ldots,n$.

PROOF. (1) Letting $N_i = \Sigma_{j \neq i=1}^{n} \Lambda x_j$, and $A_i = N_i^R$, we have by 8, that $A_i^M = N_i$. Thus $x_i \notin N_i \to x_i A_i \neq 0$ and $x_i A_j = 0$, $j \neq i = 1, \ldots, n$. (2) If $y_i \in x_i A_i$, $y_i = x_i a_i$, $a_i \in A_i$, $i=1, \ldots, n$, and $a = a_1 + \ldots + a_n$ has the required property.

M_R is an <u>irreducible</u> module in case (1) M_R is simple, i.e., 0 and M are the only submodules, and (2) $MR \neq 0$. A simple module $M \neq 0$ is irreducible if and only if

$R^M = 0$. For, if $R^M = 0$, then (2) holds. Conversely, if (2) holds, then R^M is a submodule $\neq M$, so $R^M = 0$.

10. LEMMA. If M_R is simple and $M \neq 0$, then $\Lambda = \mathrm{Hom}_R(M,M)$ is a field.

PROOF. $\Lambda \neq 0$ since the identity map $1: x \to x$ differs from the zero map $0: x \to 0$. If $\lambda \neq 0 \in \Lambda$, then λM is a nonzero submodule of M_R, so $\lambda M = M$. Thus λ is an epimorphism. Now $\mathrm{Ker}(\lambda)$ is a submodule $\neq M = \lambda M$, so $\mathrm{Ker}(\lambda) = 0$ and λ is a monomorphism. This makes λ an automorphism (= an isomorphism of M onto M), so λ^{-1} exists. Then Λ is a (not necessarily commutative) field.

Let Q denote the field of rational numbers. Let $M = (Q,+)$ denote the additive group of Q. Then $\mathrm{Hom}_Z(M,M)$ is a field isomorphic to Q, but M_Z is not a simple module.

The lemma shows that if M is an irreducible R-module, then M is a left vector space over Λ. In this case the definition of linear independence of elements of M over Λ agrees with the usual one for vector spaces M. Furthermore, since M_R is irreducible, in the terminology of 9, if $x_1, \ldots, x_n \in M$ are linearly independent over Λ, then $x_i A_i = M$, $i=1, \ldots, n$. Thus 9 establishes the next result.

11. PROPOSITION. Let M_R be irreducible, and let $\Lambda = \mathrm{Hom}_R(M,M)$. Then ${}_\Lambda M$ is a vector space. If $x_1, \ldots, x_n \in M$ are linearly independent, and if $y_1, \ldots, y_n \in M$ are arbitrary, then there exists $r \in R$ such that $x_i r = y_i$, $i=1, \ldots, n$.

DEFINITION 1. Let A be a full ring of l.t.'s in a left vector space M over a field Λ. A subring B of A is a <u>dense</u> subring of A in case B has the following property: if $x_1,\ldots,x_n$ are finitely many linearly independent elements of M, and if $y_1, \ldots, y_n \in M$, then there exists $b \in B$ such that $x_i b = y_i$, $i = 1, \ldots, n$.

DEFINITION 2. A module M_R is <u>faithful</u> in case $M^R = 0$, that is, in case $\forall\, r \in R$, $Mr = 0$ implies $r = 0$.

It is easy to see that if M_R is any module, then for each $r \in R$, the correspondence $\bar{r}: x \to xr$ which sends each $x \in M$ into xr is an element of $A = \mathrm{Hom}_\Lambda(M,M)$, where $\Lambda = \mathrm{Hom}_R(M,M)$. Furthermore, the correspondence $r \to \bar{r}$ is a ring homomorphism of R into $\bar{R}$, called the <u>representation</u> of R in M. If M_R is faithful, then the homomorphism is a monomorphism and then the representation is said to be faithful. If M_R is irreducible and faithful, then the representation is said to be irreducible and faithful.

Going back to 11, if M_R is irreducible and faithful, then R is isomorphic to a dense subring of $A = \mathrm{Hom}_\Lambda(M,M)$ under the correspondence $r \to \bar{r}$ just defined. This establishes the following:

12. COROLLARY. If M_R is a faithful, irreducible module, then R is isomorphic to a dense ring of linear transformations in $_\Lambda M$, where $\Lambda = \mathrm{Hom}_R(M,M)$.

We say a ring R is right primitive in case R has a faithful irreducible module M_R . Thus 12 states that each primitive ring is isomorphic to a dense ring of l.t.'s in some (left) vector space. The next result gives the converse.

13. PROPOSITION. If R is isomorphic to a dense ring of l.t.'s in a left vector space V over a field F , then R is a primitive ring.

PROOF. Let $A = \mathrm{Hom}_F(V,V)$, and let B be a dense subring of A such that $R \cong_\varphi B$. Then, since V is a B-module, V is an R-module where $xr = x\varphi(r)$ $\forall r \in R,\ x \in V$. Now let W be any non-zero R-submodule (= B-submodule). If $y \in V$ and if $0 \neq x \in W$, then by density of B there exists $b \in B$ such that $y = xb$. But then $y \in W$, so that $V \subseteq W$ and $V = W$. This shows that V_R is irreducible. Since V_R is also faithful, R is a primitive ring.

14. PROPOSITION. If R is a primitive ring, and if I is any nonzero ideal of R, then I is a primitive ring.

PROOF. Let M be a faithful irreducible R-module, then $MI \neq 0$. Since MI is an R-submodule of M, $MI = M$ follows. Since M is faithful for R, it is faithful for I. It remains only to show that M is a simple I-module. Let N be any I-submodule $\neq M$. Then NI is an R-submodule $\subseteq N$, and so $NI = 0$. Since I is an ideal, $I^M = \{x \in M \mid xI = 0\}$ is an R-submodule, and $I^M \supseteq N$. However, $I^M \neq M$ since M_R is faithful and $I \neq 0$. Thus $I^M = N = 0$, and M_I is simple, as required.

REMARK. Let $\Lambda = \mathrm{Hom}_R(M,M)$, and let $\Omega = \mathrm{Hom}_I(M,M)$. If $\omega \in \Omega$, and if $x \in M,\ r \in R$, then $\omega(xrI) = \omega(x)rI = \omega(xr)I$, so that $[\omega(x)r-\omega(xr)]I = 0$. Since M_I is faithful, it follows that $\omega(x)r = \omega(xr)$, that is, $\omega \in \Lambda$. Thus, $\Omega \subseteq \Lambda$, and, trivially, $\Lambda \subseteq \Omega$, so $\Lambda = \Omega$. This shows that I is dense ring of l.t.'s in the same vector space ${}_{\Lambda}M$ in which R is dense.

A ring S is simple in case $S^2 \neq 0$ and S contains no ideals other than 0 and S.

15. LEMMA. If S is a simple ring with identity element, then S is primitive.

PROOF. By Zorn's lemma, S has a maximal right ideal I. Let M denote the simple S-module S-I. Then M_S is unital, so M_S is irreducible. Since $M^S \neq S$, it follows that $M^S = 0$, that is, that M_S is faithful. Hence S is (right, also left) primitive.

E. Sasiada has announced in Bulletin de l'Académie Polon. des Sciences, 1961, that there exist simple rings which are not primitive.

A module M_R is artinian (resp. noetherian) in case M_R satisfies the minimum (resp. maximum) condition on submodules. A ring R is right artinian (resp. noetherian) in case R_R is artinian (resp. noetherian). Similarly, R is left artinian (resp. noetherian) in case ${}_RR$ is artinian (resp. noetherian).

16. LEMMA. A full ring Λ_n of n x n matrices over a field Λ is simple and both right and left artinian and noetherian.

PROOF. Let I be a nonzero ideal of $R = \Lambda_n$. Let $\{e_{ij} \mid i,j = 1,\dots,n\}$ be a set of matrix units of R, and let

$$x = \Sigma_{i,j=1}^{n} e_{ij}\lambda_{ij}, \quad \lambda_{ij} \in \Lambda,$$

be a non-zero element of I. Then $\lambda_{i_oj_o} \neq 0$, for some i_o, j_o, and then

$$y = e_{ii_o} x e_{j_oj} = e_{ij}\lambda_{i_oj_o} \in I \quad \forall\, i,j.$$

Since $\lambda_{i_oj_o} \neq 0$, we get $e_{ij} = y\lambda_{i_oj_o}^{-1} \in I \ \forall\, i,j$. Then clearly $I \supseteq \Sigma_{i,j=1}^{n} e_{ij}\Lambda = \Lambda_n$ i.e., $I = \Lambda_n$, as required.

Viewing Λ, as in the proof above, as the set of all scalar matrices of Λ_n, one sees that Λ_n is a left and right vector space over Λ, and that $\{e_{ij} \mid i,j = 1,\ldots,n\}$ is a basis. Since the left and right ideals of Λ_n are vector subspaces, and since Λ_n has finite dimension n^2 over Λ, the last statement follows.

DEFINITION. We say that R is a full linear ring in case R is isomorphic to a full ring of linear transformations in a left vector space V over a field D. If $\dim V = n < \infty$, we say that R is a finite dimensional (f.d.) full ring.

17. LEMMA. R is a f.d. full ring if and only if R is isomorphic to a full ring D_n of all $n \times n$ matrices over a field D.

PROOF. Let V be a left vector space of dimension $n < \infty$ over D, and let $u_1, \ldots, u_n$ be a basis of V. We can assume $R = \mathrm{Hom}_D(V,V)$. Then, if $a \in R$, there exist $\delta_{ij} \in D$ such that

$$u_i a = \Sigma_{j=1}^n \delta_{ij} u_j \qquad i,j = 1, \ldots, n.$$

If (δ_{ij}) denotes the matrix with δ_{ij} in the (i,j) position, then we define $\varphi(a) = (\delta_{ij})$. If $(\gamma_{ij}) \in D_n$, $\gamma_{ij} \in D$, then $u_i \to \Sigma_{j=1}^n \gamma_{ij} u_j$ defines an element of R. Thus, $\varphi(R) = D_n$. It is easy to check that φ is an isomorphism of R and D_n.

18. THEOREM. R is primitive and right artinian if and only if R is a f.d. full ring. (Then R is simple).

PROOF. Let M be a faithful irreducible R-module. Then R is isomorphic to a dense ring $\bar{R}$ of l.t.'s in the left vector space M over Λ, where $\Lambda = \mathrm{Hom}_R(M,M)$. Assume that R is right artinian, let $x_1, \ldots, x_n, \ldots$ be linearly independent vectors of M, and let $V_n = \Sigma_1^n \Lambda x_i$, n=1,2,... . Then by 8, $V_n^{RM} = V_n$, n=1,2,... . Since $V_1 \subset V_2 \subset \ldots$, this shows that $V_1^R \supset V_2^R \supset \ldots$. Since R is right artinian, this shows that the sequence $V_1^R \supset V_2^R \supset \ldots$ is finite, that is, that the set $x_1, x_2, \ldots$ is finite. Thus, $\dim M < \infty$. Let $x_1, \ldots, x_n$ be a basis of M. If $a \in \mathrm{Hom}_\Lambda(M,M)$, then by density of $\bar{R}$, there exists $\bar{r} \in \bar{R}$ such that $x_i\bar{r} = x_i a$, i=1,...,n. Then $a = \bar{r} \in \bar{R}$, so $\bar{R} = \mathrm{Hom}_\Lambda(M,M)$, and R is a f.d. full linear ring. By 17, $R \cong \Lambda_n$, so R is simple. Conversely, if R is a f.d. full linear ring, then R is a dense ring of l.t.'s in a f.d. vector space V over a field D , so R is primitive. Since R is then $\cong D_n$, where n = dim , it follows that R is right (and left) artinian (and noetherian).

The product $A \cdot B$ of two right (resp. left) ideals of a ring R is defined to be the right (resp. left) ideal generated by the set $\{ab \mid a \in A, b \in B\}$. (Then $A \cdot B$ consists of all finite sums of elements of this set.) Associativity of the operation $A \cdot B$ follows immediately from that of the ring multiplication in R . Thus the product $A_1 \ldots A_n$ of finitely many right (resp. left) ideals can be unambiguously defined inductively. We set $A^1 = A$, $A^2 = A \cdot A$, and $A^n = A^{n-1} \cdot A$. A right (resp. left) ideal A is nilpotent (of index n) in case $A^n = 0$ (and either $n = 1$ or $A^{n-1} \neq 0$). For convenience, R is said to have no nilpotent right (resp. left) ideals if 0 is the only nilpotent right (resp. left) ideal.

19. **LEMMA.** R has no nilpotent ideals if and only if R has no nilpotent right (resp. left) ideals.

PROOF. Let I be a nilpotent right ideal of index $n > 1$. Then $J = I^{n-1}$ is nilpotent of index 2. Now RJ is an ideal of R , and

$$(RJ)^2 = (RJ)(RJ) = R(JR)J \subseteq RJ^2 = 0 .$$

If $RJ \neq 0$, we are through, since RJ is an ideal. If $RJ = 0$, then $B = \{b \in R \mid Rb = 0\}$ is an ideal of R satisfying $B^2 = 0$. Since $B \supseteq J$, $B \neq 0$, and the proof is complete.

Let $\{R_i \mid i \in I\}$ be a family of rings, and let R denote the direct product of the additive groups $\{(R_i,+), i \in I\}$. We make R into a ring, denoted by $\Pi_{i \in I} R_i$, and called the <u>direct product of the rings</u> $\{R_i \mid i \in I\}$, by defining multiplication in R componentwise:

$$(\ldots, x_i, \ldots)(\ldots, y_i, \ldots) = (\ldots, x_i y_i, \ldots).$$

Then R_i is isomorphic to a subring of R under the correspondence $x_i \to (0, \ldots, x_i, \ldots, 0) \ \forall x_i \in R_i$. Identifying R_i with its image under this ring monomorphism, it is easily seen that R_i is an ideal of R , and that $R_i R_j = 0 \ \forall i \neq j \in I$. The subring $\Sigma \oplus R_i$ of $R = \Pi_{i \in I} R_i$ corresponding to the direct sum of the groups $\{(R_i,+) \mid i \in I\}$ is called the <u>direct sum of the rings</u> $\{R_i \mid i \in I\}$.

20. LEMMA. Let R be a ring, and let $\{R_i \mid i \in I\}$ be a family of subrings such that: (1) the groups $\{(R_i,+) \mid i \in I\}$ are independent; (2) $(R,+)$ is the sum $\Sigma_{i\in I}(R_i,+)$. Then the correspondence $\eta\colon (\ldots, x_i, \ldots) \to \Sigma_{i\in I} x_i$ is a ring isomorphism of $\Sigma_{i\in I}\oplus R_i$ onto R if and only if R_i is an ideal of R $\forall\, i \in I$.

PROOF. Let $P = \Sigma_{i\in I}\oplus R_i$ be the ring direct sum of the rings $\{R_i\}$. Then η is an isomorphism of $(P,+)$ onto $(R,+)$. If η is a ring isomorphism, then R_i is an ideal of R , since $\eta^{-1}R_i$ is an ideal of P, $\forall\, i \in I$. Conversely, if R_i is an ideal of R $\forall\, i \in I$, then $R_iR_j = 0$ $\forall\, i \neq j \in I$, since $R_iR_j \subseteq R_i \cap R_j$, and $R_i \cap R_j = 0$ if $i \neq j$. Then, if $x = (\ldots, x_i, \ldots)$, $y = (\ldots, y_j, \ldots) \in P$,

$$\eta(xy) = \Sigma_{i\in I} x_i y_i = (\Sigma_{i\in I} x_i)(\Sigma_{j\in I} y_j) = \eta(x)\eta(y) ,$$

so η is a ring isomorphism.

We come to the classical Wedderburn-Artin theorem.

21. THEOREM. Let R be a ring $\neq 0$. Then R is an artinian ring with no nilpotent ideals if and only if R is a direct sum of a finite number of finite dimensional full rings.

PROOF. Assume that R is artinian and has no nilpotent ideals. Let I be any minimal (nonzero) right ideal of R , and let $K = \{a \in R \mid Ia = 0\}$. Since $I^2 \neq 0$, $I \not\subseteq K$, and I is therefore an irreducible R-module. Since K is an ideal of R , I a faithful R-K module under the definition $x(r+K) = xr$ $\forall\, r \in R$, $r+K \in R-K$. Thus, R-K is a primitive ring. Since R-K is artinian (along with R), by 16, R-K is a f.d. full ring. We shall make use of the fact (18) that a f.d. full ring is simple as follows: $P = RI$ is a nonzero ideal of R and $P \not\subseteq K$, so simplicity of R-K implies that $R = P + K$. Now $D = P \cap K$ is also an ideal of R , and

$$D^2 = D\cdot D \subseteq (RI)\cdot K = R\cdot(IK) = R\cdot 0 = 0 ,$$

Since R has no nilpotent ideals, $D = 0$, so that $R = P \oplus K$ is a ring direct sum of the ideals P and K . It follows that each (right, left) ideal of K is a (right, left) ideal of R , and consequently, K has no nilpotent ideals and K is also artinian. Since $R-K \approx RI$ is a f.d. full ring, a similar ring direct sum $K = Q \oplus S$, where Q is a f.d. full ring, holds for K . By the minimum condition, this process leads in a finite number of steps to the stated decomposition of R into a ring direct sum of finitely many f.d. full rings. The converse is easily checked.

4. RADICAL AND SEMIPRIMITIVITY IN RINGS

We repeat some earlier definitions and add some new ones. If M_R is any module, then $(0:M) = \{r \in R \mid Mr = 0\}$ is an ideal of R called the annihilator of M. Then the canonical ring epimorphism $R \to R - (0:M)$ is called a representation of R (in M_R). Clearly, M is also an $R-(0:M)$ module under the definition

$$x(r+(0:M)) = xr \quad \forall\, x \in M\ ,\ r+(0:M) \in R-(0:M)\ .$$

The representation of R in M is faithful in case $(0:M) = 0$. Then M_R is called a faithful R-module. (In any event, M is always a faithful $R-(0:M)$ module under the above definitions.)

If I is an ideal of R, and if M is a $R-I$ module, then by defining

$$xr = x(r+I) \quad \forall\, x \in M,\ r \in R,$$

M becomes an R-module and the annihilator $(0:M)$ of M in R contains I. Furthermore, $(0:M) = I$ if and only if M is a faithful $R-I$ module.

A module M_R is irreducible in case M_R is simple and $MR \neq 0$. It is to be noted that the requirement that R possesses an irreducible module M_R places a restriction on R. For then, MR is a nonzero submodule of M, and so $M = MR$. By iteration, $M = MR^k$ for any natural number k; in particular $R^k \neq 0\ \forall\, k$, i.e., R can not be nilpotent. However, any ring R with identity element possesses an irreducible module. For, by Zorn's lemma, R contains a maximal right ideal I. Then $M = R-I$ is a simple, unital R-module, i.e., M is an irreducible R-module.

The radical $J(R)$ of R is defined to be the intersection of the kernels of all irreducible representations of R in case R has an irreducible representation, and $J(R)$ is defined $= R$ otherwise. If $\vartheta(R)$ denotes the totality of irreducible R-modules, and if $\vartheta(R)$ is non-empty, then according to the definition:

$$J(R) = \bigcap_{M \in \vartheta(R)} (0:M)\ ,$$

R is primitive in case $(0:M) = 0$ for some $M \in \vartheta(R)$ (see 3). For reasons to be explained later, R is said to be <u>semiprimitive</u> in case $J(R) = 0$. A semiprimitive ring, then, has the property that given $0 \neq r \in R$ there exists some irreducible module M_R such that $Mr \neq 0$. Then, if $a, b \in R$, and if $a \neq b$, then there exists $M \in \mathfrak{P}(R)$ and $x \in M$ such that $xa \neq xb$. Expressed otherwise, R is semiprimitive if and only if R has enough irreducible representations in order to distinguish elements of R .

A module M_R is <u>strictly cyclic</u> if and only if there exists $u \in M$ such that $M = uR$. The element u is called a generator of M . (In general a module M is <u>cyclic</u> in case M is generated by a single element, i.e., $M = xR+xZ$ for some $x \in M$.)

A right ideal I of R is called <u>modular</u> if and only if there exists $e \in R$ such that $a-ea \in I \ \forall\, a \in R$. The element e is called a <u>left identity modulo</u> I.

Note, if R has an identity element, then each right ideal of R is modular, and if M_R is unital, then a submodule N is strictly cyclic with generator u if and only if N is the submodule of M generated by u .

1. PROPOSITION. M_R is strictly cyclic if and only if $M \cong R-I$ where I is a modular right ideal of R . A right ideal I of R is modular if and only if $I = (0:u)$ where u is the generator of a strictly cyclic R-module.

PROOF. Let $M = uR$ be strictly cyclic with generator u . The correspondence $a \to ua \ \forall\, a \in R$ is an epimorphism $R_R \to M_R$ with kernel $I = (0:u)$. Thus $M \cong R-I$. Writing $u = ue$ for some $e \in R$, we see that $a-ea \in I = (0:u) \ \forall\, a \in R$ so that I is modular with left identity e .

Conversely, if I is modular with left identity e , then $M = R-I$ is strictly cyclic with generator $e+I$, since $(e+I)a = a+I \ \forall\, a \in R$, i.e., $M = uR$, where $u = e+I$.

2. COROLLARY. If I is a modular right ideal, then $I \supseteq (I:R)$, where

$$(I:R) = \{a \in R \mid Ra \subseteq I\} .$$

PROOF. If $a \in (I:R)$, and if e is the left identity modulo I, then $ea \in I$, and $a-ea \in I$, whence $a \in I$.

NOTE. If I is modular, then $(I:R)$ is the kernel of the representation of R in $R-I$. Thus, $(I:R)$ is an ideal in R . Furthermore, $(I:R)$ contains each ideal of

$R \subseteq I$, for if J is an ideal of R, and if $J \subseteq I$, then $RJ \subseteq J \subseteq I$, so that $J \subseteq (I:R)$. Hence $(I:R)$ is the largest ideal of R contained in I.

Following Jacobson [1, Chap. 1], we give two characterizations of the irreducible modules of R. It turns out that the irreducible R-modules coincide (in the sense of module isomorphism) with the modules R-I, where I ranges over all modular maximal right ideals of R. This characterization is valuable because it provides an intrinsic characterization of the irreducible representations of R and their kernels.

We recall that if M_R is an R-module, then $R^M = \{x \in M \mid xR = 0\}$.

3. PROPOSITION. M_R is irreducible if and only if (1) $M \neq 0$ and (2) M is strictly cyclic with each nonzero element of M as a generator.

PROOF. If M is irreducible then $MR \neq 0$ so $M \neq 0$. Since R^M is a submodule of M, and $R^M \neq M$, then $R^M = 0$. Thus, if $0 \neq u \in M$, $uR \neq 0$, so $M = uR$. Conversely, (2) implies that $M = uR$, if $0 \neq u \in M$, and then $M = MR \neq 0$. If N is any nonzero submodule, and if $0 \neq r \in N$, then (2) implies $M = rR \subseteq N$, so $M = N$, and M is irreducible.

4. PROPOSITION. M_R is irreducible if and only if $M \cong R\text{-}I$, where I is a modular maximal ideal.

PROOF. If M is irreducible, and if $0 \neq u \in M$, then $M = uR$ by 3. Then, by 1, $M \cong R\text{-}I$, where $I = (0:u)$ is a modular right ideal. I is a maximal right ideal because of the obvious 1-1 correspondence between right ideals of R containing I and the R-submodules of $M = R\text{-}I$. Conversely, if I is a modular maximal right ideal, then $M = R\text{-}I$ is a simple R-module (since I is maximal). Furthermore, $MR = 0 \rightarrow R \subseteq (I:R) \subset R$ (see the note following 2.), a contradiction. It follows that $M = R\text{-}I$ is irreducible.

An ideal B of R is <u>primitive</u> in case B is the kernel of some irreducible representation of R, that is, in case $B = (0:M)$ for some irreducible R-module M, or equivalently, in case R-B is a primitive ring. Identifying isomorphic R-modules, by 4, we see that

$$\vartheta(R) = \{R\text{-}I \mid I \text{ a modular maximal right ideal of } R\}.$$

However, if $M = R\text{-}I$, then $(0:M) = (I:R)$, and we have:

5. COROLLARY. The set of primitive ideals of R coincides with

$$\{(I:R) \mid I \text{ modular maximal right ideal of } R\} .$$

6. THEOREM. $J(R)$ is the intersection of the modular maximal right ideals of R.

PROOF. By definition, $J(R)$ is the intersection of the primitive ideals of R, and by 5, these are the ideals $(I:R)$, where I ranges over the modular maximal right ideals. Since $I \supseteq (I:R)$ by 2, we conclude that $J(R)$ is contained in the intersection of the modular maximal right ideals. On the other hand, $J(R) = \cap (O:M)$, where M ranges over the irreducible R-modules. Now, for any such M ,

$$(O:M) = \bigcap_{O \neq u \in M} (O:u) ,$$

and the proof of 4 shows that $(O:u)$ is a modular maximal right ideal for each $O \neq u \in M$. This shows that $J(R)$ contains the intersection of the modular maximal right ideals, proving the theorem.

An element $x \in R$ is <u>right quasi-regular</u> (r.q.r.) in case there exists an element $x' \in R$ such that $x+x'+xx' = O$. Then x' is called a <u>r.q.inverse</u> of x. In case R has an identity element 1, this is equivalent to the requirement of an element $x' \in R$ such that $(1+x)(1+x') = 1$. (In this case, x is r.q.r. if and only if $1+x$ has a right inverse y in R , and then $x' = y-1$ is a r.q.inverse of x). An element $x \in R$ is left quasi-regular (l.q.r.) in case there exists $x'' \in R$ such that $x+x''+x''x = O$, and then x'' is called a l.q.inverse of x . In the same way as the case of inverses in rings with identity, we see that if x has both a r.q.inverse x' and a left quasi-inverse x'', then $x' = x''$. In this case x is said to be <u>quasi-regular</u> (q.r.).

If $u \in R$ is nilpotent, that is, if $u^k = O$ for some natural number k , then $u' = -u+u^2-u^3+\dots(-1)^{k-1}u^{k-1}$ is an element of R , and, in fact, u' is both a left and right quasi-inverse of u . (If $1 \in R$, this is equivalent to the evident assertion that $(1+u)^{-1} = 1-u+u^2-u^3+\dots(-1)^{k-1}u^{k-1}$. Thus, each nilpotent $u \in R$ is q.r.

We say that a one-sided ideal I of R is <u>quasi-regular</u> (q.r.) in case each element of I is q.r. Our remark above establishes that each nil one-sided ideal (= a one-sided ideal such that each element is nilpotent) is q.r.

7. LEMMA. A right (resp. left) ideal I is q.r. if and only if each $x \in I$ is r.q.r. (resp. l.q.r.).

PROOF. If $x \in I$, and if x has a r.q. inverse x', then $x' = -x-xx' \in I$, so that x' has a r.q. inverse x''. Then we compute:

$$x = x+0+x\cdot 0 = x+(x'+x''+x'x'') + x(x'+x''+x'x'')$$
$$= x''+(x+x'+xx') + (x+x'+xx')x'' ;$$

and

$$x'' = x''+0+0\cdot x'' = x'' + (x+x'+xx') + (x+x'+xx')x'' ;$$

so that $x = x''$ has x' as left quasi-inverse. This shows that x is q.r. and so is I. The converse is trivial.

By symmetry we define the <u>left radical</u> $J'(R)$ to be the intersection of the ideals $(0:M')$, where M' ranges over all irreducible left R-modules. An ideal I or R is a left primitive ideal in case $I = (0:M')$ for some irreducible R-module M', and R is a <u>left primitive ring</u> if 0 is a left primitive ideal of R.

8. THEOREM. (1) $J(R)$ is a quasi-regular ideal containing each quasi-regular right ideal of R. (2) $J(R)$ coincides with the left radical $J'(R)$.

PROOF. $J(R)$ is an ideal since it is defined as the intersection of ideals. For each $x \in R$ the set $\{a+xa \mid a \in R\}$ is a right ideal. Moreover, x is r.q.r. $\leftrightarrow x \in \{a+xa \mid a \in R\}$. Whether or not R has an identity element, denote $\{x+xa \mid a \in R\}$ by $(1+x)R$. If x is not r.q.r., then $x \notin (1+x)R$, and by Zorn's lemma there exists a right ideal I_x of R which is maximal in the set of all right ideals which contain $(1+x)R$ and which do not contain x. But then it is easy to see that I_x is a maximal right ideal, since any right ideal $Q \supset I_x$ contains x, hence contains $a = (a+xa)-xa \quad \forall a \in R$. Furthermore, I_x is a modular maximal right ideal, since $-x$ is a left identity modulo I_x. This shows that any x belonging to the intersection of the modular maximal right ideals of R must be r.q.r. By 6, (and 7), $J(R)$ is q.r.

Let H be any q.r. right ideal, and assume that $H \not\subseteq J(R)$. Since $J(R) = \cap (I:R)$, I ranging over all modular maximal right ideals, we conclude that there exists some modular maximal right ideal I such that $H \not\subseteq (I:R)$. Now $R-I$ is an irreducible R-module with generator e (= any left identity modulo I). Thus, $(e+I)H = R-I$, consequently there exists $x \in H$ such that $(e+I)x = -e+I$, that is, such that $y = e+ex \in I$. But if x' is a quasi-inverse of x, then

$$y+yx' = (e+ex)+(ex'+exx') = e+e(x+x'+xx') = e \in I ,$$

which is a contradiction. This establishes (1).

(2) By symmetry $J'(R)$ is a quasi-regular ideal containing each quasi-regular left ideal. Thus $J'(R) \supseteq J(R)$, and equality follows by symmetry.

REMARK. By 6, we conclude that the intersection of the modular maximal right ideals coincides with the intersection of the modular maximal left ideals of R.

9. COROLLARY. For any ring R, $R-J$ is semiprimitive, where $J = J(R)$.

PROOF. Let I be an ideal of R containing J and mapping onto $J(R-J)$ under the canonical homomorphism $R \to R-J$. If $x \in I$, then since $I-J$ is a quasi-regular ideal of $R-J$, there exists $y \in R$ such that $x+y+xy \in J$. Then, since J is a quasi-regular ideal of R, there exists $w \in R$ such that

$$x+y+xy+w+(x+y+xy)w = 0 .$$

Then $x+u+xu = 0$, where $u = y+w+yw$, so that x is a quasi-regular element of R. This shows that I is a quasi-regular ideal of R, so that $I \subseteq J$. Since $I \supseteq J$, we see that $I = J$. Consequently $J(R-J) = 0$, and $R-J$ is semiprimitive.

Let $\{R_i \mid i \in I\}$ be any family of rings and let B be any subring of the ring direct product (§ 3) $R = \prod_{i \in I} R_i$. For each $i \in I$ let π_i denote the projection map $R \to R_i$. Then B is called a <u>subdirect sum</u> of R in case for each $i \in I$ and each element $r \in R_i$, there exists $b \in B$ (depending on i and r) such that $\pi_i b = r$. We also say that B is a <u>subdirect sum of the rings</u> $\{R_i \mid i \in I\}$. Note that the direct sum $\Sigma_{i \in I} \oplus R_i$ is a subdirect sum of the rings $\{R_i \mid i \in I\}$.

10. LEMMA. B is isomorphic to a subdirect sum of rings $\{R_i \mid i \in I\}$ if and only if the following two conditions are satisfied: (1) to each $i \in I$ there corresponds an ideal Q_i of B such that $B-Q_i \cong R_i$; (2) $\cap_{i \in I} Q_i = 0$.

PROOF. Let B be a subdirect sum of rings $\{R_i \mid i \in I\}$. Then the projection map $\pi_i: R \to R_i$, where $R = \Pi_{i \in I} R_i$, induces a ring epimorphism $\pi_i: B \to R_i$. Then $B-Q_i = R_i$, where $Q_i = \mathrm{Ker}(\pi_i)$. If $b \in \cap_{i \in I} Q_i$, then $\pi_i b = 0 \ \forall\, i \in I$, and then $b = 0$. The ideals Q_i therefore satisfy conditions (1) and (2). If B' is a ring isomorphic to B under a map φ , then (1) $B'-\varphi^{-1} Q_i \cong R_i$ and (2) $\cap \varphi^{-1}(Q_i) = 0$.

Conversely suppose ideals $\{Q_i \mid i \in I\}$ of B satisfy (1) and (2), and consider the mapping f of B into $R = \Pi_{i \in I} R_i$ defined as follows:

$$f: B \to (\ldots, \varphi_i(b), \ldots) \quad \forall\, b \in B ,$$

where φ_i is the homomorphism $B \to R_i$ whose kernel is $Q_i \ \forall\, i \in I$. f is clearly a ring homomorphism $f: B \to R$, and f is onto R since each φ_i is onto R_i . If $f(b) = 0$, then $b \in \bigcap_{i \in I} \mathrm{Ker}(\varphi_i) = \bigcap_{i \in I} Q_i = 0$, and $b = 0$, so that B is isomorphic (under f) to a subdirect sum $f(B)$ of $R = \Pi_{i \in I} R_i$.

11. THEOREM (Jacobson). A ring R is semiprimitive if and only if R is isomorphic to a subdirect sum of primitive rings.

PROOF. By the lemma, it is equivalent to show that $J(R) = 0$ if and only if there exist ideals Q_i of R satisfying (2), where $R-Q_i$ is a primitive ring $\forall\, i \in I$. Now if $J(R) = 0$, then $\cap\, (0:M) = 0$, where M ranges over all irreducible R-modules, and the rings $R-(0:M)$ are primitive. Then, the lemma implies the theorem. Conversely, if Q_i are ideals with the property described above, then $R-Q_i$ has a faithful irreducible module $M_i \ \forall\, i \in I$. Thus M_i is also an irreducible R-module satisfying $(0:M_i) = Q_i$. Since $J(R) \subseteq (0:M_i) = Q_i \ \forall i$, and since $\bigcap_{i \in I} Q_i = 0$, we include that $J(R) = 0$, and that R is semiprimitive.

A ring R is a <u>von Neumann ring</u>*) in case to each $a \in R$ there corresponds $x \in R$ (depending on a) such that $axa = a$. For each $a \in R$, $|a)$ denotes the right ideal

---

*) Called a <u>regular</u> ring in the literature. See the Introduction.

generated by a, and $(a|$ denotes the left ideal generated by a.

12. PROPOSITION. R is a von Neumann ring if and only if for each $a \in R$ there exists $e = e^2 \in R$ such that $|a) = eR$ (resp. $(a| = Re$).

PROOF. If R is von Neumann, then given $a \in R$ there exists $x \in R$ such that $a = axa$. Then $e = ax$ (resp. $e = xa$) satisfies $e^2 = e$, and $eR \subseteq |a)$ (resp. $Re \subseteq (a|$). But $a = ea \in eR$ (resp. $a = ae \in Re$) so that $|a) \subseteq eR$ (resp. $(a| \subseteq Re$), and the desired equality follows. Conversely assume that $|a) = eR$ with $e = e^2 \in R$. Now $|a) = aR+Za$, and $e = e^2 \in |a)$, so there exist $r \in R$, $n \in Z$ such that $e = ar + na$. $e = e^2 = (ar+na)^2 = ay$, where $y \in R$. Since $a \in eR$, and since $ex = x \ \forall x \in eR$, we see that $ea = a$. Thus $aya = ea = a$, as required. By symmetry, it follows that $(a = Re$ implies $aya = a$ for some $y \in R$.

13. COROLLARY. A von Neumann ring is semiprimitive.

PROOF. Let $a \in J(R)$, and assume R is von Neumann. Then $|a) = eR$, with $e = e^2 \in R$. Since $|a) \subseteq J = J(R)$, also $e \in J(R)$. Let e' denote the quasi-inverse of $-e$. Then $e = e' - ee'$, and $e^2 = e = ee' - e^2e' = 0$. Thus $|a) = 0R = 0$, and $a = 0$. This shows that $J(R) = 0$, and R is semiprimitive.

We include here a property of von Neumann rings which will be used several places later on.

14. PROPOSITION. Let S be a von Neumann ring with identity, and let $P_r(S)$ denote the totality of principal right ideals of S. Then: (1) if $A,B \in P_r(S)$, then $A+B \in P_r(S)$; (2) $P_r(S)$ contains each finitely generated right ideal of S.

PROOF. (2) is an obvious corollary to (1).

(1) We have already shown that S is von Neumann if and only if each principal right ideal is generated by an idempotent. Hence, there exist $e = e^2 \in S$, $f = f^2 \in S$ such that $A = eS$ and $B = fS$. Let $B_1 = (1-e)fS$. Now $A+B = \{eu+fv \mid u, v \in S\}$, and

$$\begin{aligned} A+B_1 &= \{eu'+(1-e)fv \mid u', v \in S\} \\ &= \{e(u'-fv)+fv \mid u', v \in S\}, \end{aligned}$$

so we see that $A+B_1 = A+B$.

Write $B_1 = f_1S$, with $f_1^2 = f_1 \in S$. Since $f_1 \in B_1$, then $ef_1 = 0$. Now put $f' = f_1(1-e)$. Then

$$f'f_1 = f_1(1-e)f_1 = f_1(f_1-ef_1) = f_1f_1 = f_1 ,$$

so $(f')^2 = f'f_1(1-e) = f_1(1-e) = f'$. Since $f' = f_1(1-e) \in f_1S$, and since $f_1 = f'f_1 \in f'S$, it follows that $f'S = f_1S = B_1$. Hence $A+B = eS+f'S$. Now

$$ef' = ef_1(1-e) = 0, \quad f'e = f_1(1-e)e = 0 ,$$

so

$$A+B = eS+f'S = (e+f')S \in P_r(S) .$$

5. THE ENDOMORPHISM RING OF A QUASI-INJECTIVE MODULE

As before $M \,'\supseteq N$ denotes that N is an essential submodule of M, and $J(\Lambda)$ denotes the Jacobson radical of a ring Λ.

1. THEOREM. Let M_R be a quasi-injective module, let $\Lambda = \mathrm{Hom}_R(M,M)$, and let $J = J(\Lambda)$. Then: (1)

$$J = \{\lambda \in \Lambda \mid M \,'\supseteq \ker(\lambda)\} ,$$

and $\Lambda - J$ is a von Neumann ring; (2) If $J = 0$, then Λ_Λ is injective (and Λ is a von Neumann ring).

PROOF. Let $I = \{\lambda \in \Lambda \mid M \,'\supseteq \ker(\lambda)\}$. If $\lambda \in \Lambda, \mu, \nu \in I$, then

$$\ker(\mu+\nu) \supseteq \ker(\mu) \cap \ker(\nu),\ \ker(\lambda\mu) \supseteq \ker(\mu) .$$

Since $\ker(\mu) \cap \ker(\nu)$ is an essential submodule of M_R, it follows that I is a left ideal of Λ. Furthermore, since $\ker(\lambda) \cap \ker(1+\lambda) = 0 \ \forall \lambda \in \Lambda$, if $\lambda \in I$ it follows that $\ker(1+\lambda) = 0$. Thus, $(1+\lambda)$ has a left inverse $\forall \lambda \in I$, so that each $\lambda \in I$ is left quasi-regular. Thus, I is a left-quasi-regular left ideal, so $I \subseteq J$.

Next let λ be an arbitrary element of Λ, let L be a relative complement submodule corresponding to $K = \ker(\lambda)$, and consider the correspondence $\lambda x \to x \ \forall x \in L$. If $\lambda x = \lambda y$, with $x,y \in L$, then $\lambda(x-y) = 0$, so that $x-y \in K \cap L = 0$, that is, $\lambda x = \lambda y \leftrightarrow x = y$. Since M_R is quasi-injective the map $\lambda x \to x$ of λL in L is induced by some $\theta \in \Lambda$. If $u = x+y \in L \oplus K$, $x \in L$, $y \in K$, then

$$(\lambda - \lambda\theta\lambda)(u) = \lambda(x) - \lambda\theta\lambda(x) = \lambda(x) - \lambda(x) = 0 .$$

Since $M_R \,'\supseteq L \oplus K$, and since $\mathrm{Ker}(\lambda - \lambda\theta\lambda) \supseteq L \oplus K$, we conclude that $\lambda - \lambda\theta\lambda \in I$. Although we have not yet shown that I is an ideal (it would not be too hard to do so), roughly speaking we have shown that Λ is a von Neumann ring modulo I.

Now to show that $J = I$. If $\lambda \in J$, and if $\theta \in \Lambda$ is chosen so that $\mu = \lambda - \lambda\theta\lambda \in I$, then $-\lambda\theta \in J$ (since J is an ideal) and $(1-\lambda\theta)^{-1}$ exists (since J is quasi-regular). Therefore $(1-\lambda\theta)^{-1}\mu = \lambda$, and $\lambda \in I$ (since I is a left ideal). Thus $J = I$

as asserted. Moreover $\Lambda - J$ is von Neumann.

There remains to show that $J = 0$ implies that Λ_Λ is injective. From what already has been proved, we know that Λ is a von Neumann ring, and we will use the following property (4.14): each finitely generated right ideal of Λ is generated by an idempotent.

Let $f: I \to \Lambda$ be any map of a right ideal I into Λ. By IM we mean that submodule of M_R generated by $\{\lambda m \mid \lambda \in I, m \in M\}$, and it follows that if $x \in IM$ then there exist $\lambda_1, \ldots, \lambda_n \in I$ and $m_1, \ldots, m_n \in M$, $n \in Z^+$, such that $x = \Sigma_{i=1}^n \lambda_i m_i$. We consider a correspondence

$$x = \Sigma_{i=1}^n \lambda_i m_i \to \Sigma_{i=1}^n f(\lambda_i) m_i ;$$

if also $y = \Sigma_{j=1}^t \mu_j m_j' \in IM$, $\mu_j \in I$, $m_j' \in M$, $j = 1, \ldots, t$, then the right ideal generated by $\lambda_1, \ldots, \lambda_n, \mu_1, \ldots, \mu_t$ has the form $e\Lambda$, where $e = e^2 \in \Lambda$, and then $e\lambda_i = \lambda_i$, $e\mu_j = \mu_j$, $f(\lambda_i) = f(e)\lambda_i$, $f(\mu_j) = f(e)\mu_j$, $i = 1, \ldots, n$, $j = 1, \ldots, t$. Consequently,

$$\Sigma_{i=1}^n f(\lambda_i) m_i = \Sigma_{i=1}^n f(e)\lambda_i m_i = f(e)\ \Sigma_{i=1}^n \lambda_i m_i = f(e)x ,$$

similarly,

$$\Sigma_{j=1}^t f(\mu_j) m_j' = f(e)y$$

so that $x \to \Sigma_{i=1}^n f(\lambda_i) m_i$ is a single-valued correspondence, which we denote by θ. Furthermore, our remarks show that

$$\theta(x+y) = f(e)(x+y) = f(e)x + f(e)y = \theta(x) + \theta(y) ,$$

and if $r \in R$, then $\theta(xr) = f(e)(xr) = (f(e)x)r = \theta(x)r$, so we conclude that θ is a map of IM in M. By quasi-injectivity, θ is induced by an element of Λ, which we also denote by θ. Then,

$$(\theta\lambda)(m) = \theta(\lambda m) = f(\lambda)(m) , \quad \forall\ \lambda \in I,\ m \in M ,$$

so that $f(\lambda) = \theta\lambda \ \ \forall\ \lambda \in I$. This establishes that Λ satisfies Baer's condition (§1), and, by Baer's criterion for unital modules, we conclude that Λ_Λ is injective.

The proof of (2) is simply a rewriting of the proof of Johnson and Wong for the case M_R is injective. (1) is a similar generalization of Utumi's result for injective modules.

2. PROPOSITION. Assume the notation of Theorem 5.1, let $E = E(M_R)$, and $\Omega = \mathrm{Hom}_R(E,E)$. Then $\Omega M \subseteq M$, that is, each $\omega \in \Omega$ induces $\bar{\omega} \in \Lambda$, and $\omega \to \bar{\omega}$ is a ring epimorphism of Ω on Λ such that $J(\Omega) - f^{-1}J(\Lambda)$.

PROOF. $\Omega M \subseteq M$ by the theorem of Johnson-Wong stated in §3. Hence, $\omega \in \Omega$ induces $\bar{\omega} \in \Lambda$. Since E_R is injective, the correspondence $f: \omega \to \bar{\omega}$ is onto Λ. Thus, f is a ring epimorphism of Ω and Λ. That f maps $J(\Omega)$ onto $J(\Lambda)$ follows immediately from the characterization of $J(\Omega)$ (resp. $J(\Lambda)$) in Theorem 5.1, i.e., $J(\Omega) \subseteq f^{-1} J(\Lambda)$. If $\lambda \in f^{-1} J(\Lambda)$, then $M' \supseteq \mathrm{Ker}\, f(\lambda)$, and then $E' \supseteq \mathrm{Ker}\, \lambda$, so $\lambda \in J(\Omega)$, proving 2.

1. leads to another characterization of $J(R)$ in case R_R is injective and R has an identity. Before giving this, however, it is convenient to introduce a new notion. For any module M_R, let $Z(M_R) = \{x \in M \mid (0:x)$ is an essential right ideal of $R\}$. We recall that $(0:x) = \{r \in R \mid xr = 0\}$. More generally, if S and T are either both subsets of R or of M, then

$$(S:T) = \{r \in R \mid Tr \subseteq S\} ;$$

if S is a submodule of M (resp. right ideal of R), then $(S:T)$ is a right ideal. In the case $S = 0$ and $T \subseteq M$ (resp. $T \subseteq R$) we previously have denoted $(0:T)$ by T^R (resp. by T^r).

We plan to show that $Z(M_R)$ is a submodule of M_R, called the <u>singular submodule of M_R</u>. To do this we need (1) of the following lemma. A right ideal I of R is essential in case I_R is an essential submodule of R_R.

3. LEMMA. (1) If I is an essential right ideal of R, then $(I:a)$ is an essential right ideal of R $\forall a \in R$. (2) If N is an essential submodule of M_R, then $(N:x)$ is an essential right ideal of R $\forall x \in M$.

PROOF. (1) is a special case of (2). Let K be a nonzero right ideal of R.

(i) $xK = 0$ implies $K \subseteq (N:x) \cap K$; (ii) $xK \neq 0$ implies $N \cap xK \neq 0$; then if $0 \neq xk \in N \cap xK$, $k \in K$, then $k \in (N:x) \cap K$. This shows that $(N:x)$ meets each nonzero right ideal of R , i.e., $(N:x)$ is essential.

4. PROPOSITION. (1) $Z(M_R)$ is a submodule of M_R ; (2) $Z(R_R)$ is an ideal of R .

PROOF. If $x,y \in Z(M_R)$, and if $r \in R$, then

$$(0:x-y) \supseteq (0:x) \cap (0:y)$$

and

$$(0:xr) \supseteq ((0:x):r) .$$

Since $(0:x) \cap (0:y)$ is an essential right ideal, and since $((0:x):r)$ is an essential right ideal by (1) of the lemma, we conclude that $x-y$, $xr \in Z(M_R)$ proving (1).

(2) $Z(R_R)$ is a right ideal by (1). If $x \in Z_r(R)$ and if $a \in R$, then $(0:ax) \supseteq (0:x)$, so $ax \in Z(R_R)$. This shows that $Z(R_R)$ is an ideal.

We call $Z(R_R)$ the <u>right singular ideal of</u> R , and denote it by $Z_r(R)$.

5. COROLLARY. If R is a ring with identity, and if R_R is injective, then (1) $J(R) = Z_r(R)$. Furthermore, (2) if I is any finitely generated left ideal of R, then $I^{rl} = I$.

PROOF. (1) As we have shown before, if x_L denotes the left multiplication $x_L(a) = xa \ \forall a \in R$, x a fixed element of R , then $\Lambda = \mathrm{Hom}_R(R,R) = \{x_L \mid x \in R\}$, and, $R \cong \Lambda$ under $x \to x_L$. By 1,

$$J(\Lambda) = \{x_L \in \Lambda \mid \mathrm{Ker}(x_L) \text{ is essential in } R_R\} ,$$

and this means that

$$J(R) = \{x \in R \mid x^r \text{ is an essential right ideal}\} ,$$

i.e., $J(R) = Z_r(R)$. (2) follows immediately from 3.8.

NOTE. No added generality results from assuming R_R is only quasi-injective, since Baer's criterion shows that quasi-injectivity and injectivity are equivalent for rings with identity.

As we have noted, $Z_r(R)$ is defined for any ring R, and, in some cases (e.g., R is right self-injective with identity element), $J(R) = Z_r(R)$. However, $J(R)$ need not equal $Z_r(R)$. For if R denotes the ring of formal power series in a single variable x over a commutative field, then one can show that $J(R) = xR$, whereas $Z_r(R) = 0$, since R is an integral domain. There remains, however, the question whether $J(R) \supseteq Z_r(R)$ for each ring R. Recently Barbara Osofsky has constructed an example of a ring R satisfying $J(R) = 0$ and $Z_r(R) \neq 0$. (See 1963 Notices of the American Mathematical Society).

If R is any ring, an arbitrary collection of idempotents $\{e_i \mid i \in I\}$ of R is said to be orthogonal, in case $e_i e_j = 0$ for all $i \neq j \in I$.

6. THEOREM. Let M_R be quasi-injective, let $\Lambda = \mathrm{Hom}_R(M,M)$, and let $J = J(\Lambda)$. If $u_1, \ldots, u_t$ are finitely many orthogonal idempotents of $\Lambda - J$, then there exist orthogonal idempotents $e_1, \ldots, e_t$ of Λ such that e_i maps onto u_i under the canonical homomorphism $\Lambda \to \Lambda - J$.

PROOF. Since $u_1, \ldots, u_t, 1 - \Sigma_{i=1}^t u_i$, are also orthogonal idempotents whose sum = 1, we may as well assume that $\Sigma_{i=1}^t u_i = 1$, and $t \geq 2$. Let $\lambda_i \in \Lambda$ be such that $\lambda_i \to u_i$ under $\Lambda \to \Lambda - J$, $i = 1, \ldots, t$. Then

$$1-(\lambda_1+\ldots+\lambda_t),\quad \lambda_i-\lambda_i^2,\quad \lambda_i\lambda_j,\quad i \neq j,\ i,j = 1, \ldots, t,$$

are elements of J, and by 1, the kernels of these elements are essential submodules of M, and

$$C = \mathrm{Ker}(1-[\lambda_1+\ldots+\lambda_t]) \cap \bigcap_{i=1}^{t} \mathrm{Ker}(\lambda_i-\lambda_i^2) \bigcap_{i\neq j=1}^{t} \mathrm{Ker}(\lambda_i\lambda_j)$$

is therefore an essential submodule. We first show that the sum $S = \Sigma_{i=1}^t \lambda_i C$ is direct: Let $y = \Sigma_1^t \lambda_i x_i \in \Sigma_1^t \lambda_i C$, $x_i \in C$, $i=1, \ldots, t$. Then

$$\lambda_j y = \Sigma_1^t(\lambda_j\lambda_i)x_i = \lambda_j^2 x_j = \lambda_j x_j \quad ,$$

and therefore, $y = 0 \leftrightarrow \lambda_j x_j = 0 \ \forall_j$, so that $S = \Sigma_1^t \oplus \lambda_i C$.

We now assume that M_R is injective, and let C_i denote any injective hull of $\lambda_i C$ in M , i=1, ..., t. Then $\{C_i \ i=1, ..., t\}$ are independent submodules of M by 3, and their sum S is injective. Since $(1-\Sigma_1^t \lambda_i)c = 0 \ \forall c \in C$, that is, since $c = \Sigma_1^t \lambda_i c \ \forall c \in C$, it follows that $C \subseteq S = \Sigma_1^t C_i$, and S is therefore an essential submodule of M . Since S is a direct summand of M , we conclude that $S = M$.

Now let e_i be the projection of M on C_i , i=1, ..., t. If $c \in C$, then $c = \Sigma_1^t \lambda_i c$, and then $e_i c = \lambda_i c$. This implies that $\ker(e_i - \lambda_i) \supseteq C$, i=1, ..., t . Since C is an essential submodule, it follows that $e_i - \lambda_i \in J$, and then $e_i \to u_i$ under $\Lambda \to \Lambda - J$, i=1, ..., t. Since projection idempotents are obviously orthogonal, the proof is complete in the case M is injective.

Now let M be quasi-injective, let $E = E(M_R)$, and let $\Omega = \mathrm{Hom}_R(E,E)$. By 2, there exists a ring epimorphism $f: \Omega \to \Lambda$ which maps $J(\Omega)$ on $J(\Lambda)$, such that $J(\Omega) = F^{-1}J(\Lambda)$. Thus f induces a ring isomorphism $\bar{f}: \Omega - J(\Omega) \to \Lambda - J(\Lambda)$. Since $v_i = \bar{f}^{-1}(u_i)$, i=1, ...,t are orthogonal idempotents of $\Omega - J(\Omega)$, by the result above, there exist orthogonal idempotents $f_i \in \Omega$ such that $f_i \to v_i$, i=1, ..., t, under $\Omega \to \Omega - J(\Omega)$. Then clearly $f(f_i) = e_i$, i=1, ..., t, are orthogonal idempotents of Λ , and $e_i \to u_i$, i=1, ..., t under $\Lambda \to \Lambda - J(\Lambda)$. This completes the proof.

The above proof is due to S.U. Chase.

The corresponding result for rings (cf. 4) is the following:

7. COROLLARY. If R is any right self-injective ring with identity, then finitely many orthogonal idempotents of $R - J$, $J = J(R)$, can be "lifted" to orthogonal idempotents of R .

By "lifted", we mean that R has the property of the theorem, with R substituted for Λ.

Let M_R be any module. Then M_R is indecomposable in case any representation $M = N \oplus P$, where N and P are submodules, implies either $N = 0$ or $P = 0$. Our chief examples of indecomposable modules up to now are the simple modules. The cyclic groups of prime power $p^n > p$, or infinite order, are clearly indecomposable Z-modules which are not simple. Furthermore, let R be any <u>local ring</u>, that is, a ring with identity such that $R-J(R)$ is a field. Then R_R (also ${}_RR$) is indecomposable. For let $R = I \oplus K$, where I and K are right ideals. Then, $1 = e+f$, where $e = e^2 \in I$, and $f = f^2 \in K$. If $e \neq 0$, then (as we have shown in the proof that von Neumann rings are semiprimitive), $e \notin J(R)$, consequently e has an inverse modulo $J(R)$, since $R-J(R)$ is a field. Let v be such that $ev = 1 + j$, where $j \in J(R)$. Since $ev \in I$, and since j is quasi-regular, $ev(1+j)^{-1} = 1 \in I$, so that $I = R$, and $K = 0$. This proves that any local ring R is indecomposable as a right (or left) R-module. The next result gives a connection between indecomposability of a quasi-injective module M_R and its endomorphism ring.

8. PROPOSITION. Let M_R be quasi-injective, and let $\Lambda = \mathrm{Hom}_R(M,M)$. Then M_R is indecomposable if and only if Λ is a local ring.

PROOF. Already, from 1, we know that $\Lambda - J$, $J = J(\Lambda)$, is a von Neumann ring. Furthermore, if $M = M_1 \oplus M_2$ for submodules M_1, M_2, then the projections $\pi_i: M \to M_i$. $i=1, 2$, are orthogonal idempotents of Λ whose sum $\pi_1 + \pi_2 = 1$. Conversely, if e is any idempotent of Λ, then $M = eM \oplus (1-e)M$. Thus, we see that M is indecomposable if and only if Λ has no idempotents $\neq 0,1$. By 6, this is equivalent to the assertion that $\Lambda - J$ has no idempotents $\neq 0,1$. However, since $\Lambda - J$ is von Neumann, each principal right ideal is generated by an idempotent, a fact which shows that our condition is equivalent to the assertion that $\Lambda - J$ has no right ideals $\neq 0$, $\Lambda - J$. This latter condition is easily seen to be equivalent to the requirement that $\Lambda - J$ is a field.

6. NOETHERIAN, ARTINIAN, AND SEMISIMPLE MODULES AND RINGS

A module M_R is <u>noetherian</u> (resp. <u>artinian</u>) if for each ascending sequence $S_1 \subseteq S_2 \dots \subseteq S_n \subseteq \dots$ (resp. each descending sequence $S_1 \supseteq S_2 \dots \supseteq S_n \supseteq \dots$) of submodules, there corresponds k such that $S_k = S_{k+1} = \dots = S_{k+i}$ $\forall\ i > 0$. A module M_R satisfies the <u>maximum</u> (resp. <u>minimum</u>) condition if each non-empty collection $\{S_i \mid i \in I\}$ of submodules contains a maximal (resp. minimal) element S. It is elementary to verify that M_R is noetherian (resp. artinian) if and only if M_R satisfies the maximum (resp. minimum) condition. Noetherian modules also can be characterized as those modules with the property that each submodule is finitely generated. Observe that if $x_i \in M_R$, $i=1, \dots, t$, then

$$\Sigma_1^t x_i R + \Sigma_1^t x_i Z = \left\{\Sigma_1^t x_i r_i + \Sigma_1^t x_i n_i \mid r_i \in R,\ n_i \in Z\right\}$$

is the submodule generated by $x_1, \dots, x_t$. We denote this submodule by $(x_1,\dots,x_t)$. (In the event M_R is unital, note that $(x_1,\dots,x_t) = \Sigma_1^t x_i R$.) A ring R is right [resp. left] noetherian (resp. artinian) in case R_R [resp. ${}_R R$] is noetherian (resp. artinian).

1. LEMMA. Let N_R be a module and let P be a submodule. (1) Then N_R is noetherian (resp. artinian) if and only if $M = N-P$ and P are both noetherian (resp. artinian). (2) Each epimorph of a noetherian (resp. artinian) module is noetherian (resp. artinian).

PROOF. If N is noetherian (resp. artinian) so is each submodule P. Since there is a 1-1 correspondence between submodules of $N-P$ and submodules of N containing P, we see that $N-P$ is noetherian (resp. artinian). Since each epimorph of N is isomorphic to $N-P$ for some P, (2) follows.

Now suppose $N-P$ and P are noetherian, and let $S_1 \subseteq S_2 \subseteq \dots \subseteq S_n \subseteq \dots$ be any ascending sequence of submodules of N. Then $S_i + P$ (resp. $S_i \cap P$) is an ascending sequence of submodules containing P (resp. of P). Accordingly (*) there exists an integer k such that $S_i + P = S_k + P$ <u>and</u> $S_i \cap P = S_k \cap P$ $\forall\ i \geq k$.

For any three submodules A, B, C such that $A \supseteq B$, it is trivial to check that $A \cap (B+C) = B+(A \cap C)$. Thus

$$S_i \cap (S_k+P) = S_k+(S_i \cap P) = S_k+(S_k \cap P) = S_k \quad \forall\, i \geq k .$$

However, since $S_i+P = S_k+P \ \ \forall\, i \geq k$, it follows that $S_i \subseteq S_k+P$, that is, $S_i \cap (S_k+P) = S_i \ \ \forall\, i \geq k$. Hence, $S_i = S_k \ \ \forall\, i \geq k$, so N is noetherian.

In case N-P and P are artinian we start with a descending sequence $S_1 \supseteq S_2 \supseteq \dots \supseteq S_n \supseteq \dots$ of submodules of N, and in the same way arrive at the point (*) above, from which we conclude that N is artinian. This proves (1).

As an application of 1, we note the following relation between R and R_1 (see §1).

2. PROPOSITION. A ring R is right noetherian if and only if R_1 is.

PROOF. Since each right ideal of R is a right ideal of R_1, the sufficiency is clear. Conversely, assume that R is right noetherian. Now R is an ideal of R_1, and the difference R-module $M = R_1-R$ consists of the distinct cosets $(0,n)+R \ \ \forall\, n \in Z$, that is, $n \to (0,n)+R$ is an isomorphism of $(Z,+)$ and $(M,+)$. Since $(0,n)(r,0) = (nr,0) \in R \ \ \forall\, (0,r) \in R,\ n \in Z$, it follows that $MR = 0$, so the R-submodules of M are simply the additive subgroups of M. Since $(M,+) \cong (Z,+)$, it follows that $M = R_1-R$ is a noetherian R-module. Since R is noetherian, from 1, we deduce that R_1 is a noetherian R-module. Since each right ideal of R_1 is an R-module, R_1 is itself noetherian.

3. COROLLARY. If M is a direct sum of finitely many modules M_i, $i=1,2,\dots,n$, then M is noetherian (resp. artinian) if and only if M_i is noetherian (resp. artinian) $i=1,2,\dots,n$.

PROOF. The corollary is trivial if $n=1$; assume its validity in case M is a direct sum of $t < n$ modules. Then $P = \Sigma \oplus_1^{n-1} M_i$ is noetherian (resp. artinian) if and only if M_i, $i=1,\dots,n-1$ is noetherian (resp. artinian). By the theorem, M is noetherian (resp. artinian) if and only if both M-P and P are noetherian (resp. artinian). Since $M-P \cong M_n$, we conclude that M is noetherian (resp. artinian) if and only if M_i , $i=1,\ldots,n$, is noetherian (resp. artinian).

4. COROLLARY. If R is artinian (resp. noetherian) each finitely generated unital module M_R is artinian (resp. noetherian).

PROOF. Let $M_R = \Sigma_1^n x_i R$, and let F_R denote the free module on n generators. Then, since F_R is isomorphic to the direct sum of n copies of R_R , by 2, F_R is artinian (resp. noetherian). Since M_R is an epimorph of F_R , M_R is artinian (resp. noetherian) by (2) of 1.

5. PROPOSITION. For any ring R the following statements are equivalent:

(A) Each direct sum of injective modules is injective.

(B) R_R is noetherian.

PROOF. Prop. 2 states that R is right noetherian if and only if R_1 is. Furthermore (§1), M_R is injective if and only if M_{R_1} is u-injective. Taken together, these two statements show that the proposition is equivalent to its special case where R has an identity and each module in (A) is unital.

First assume only that each direct sum of countably many injective unital modules is <u>quasi-injective</u>. If $0 = I_0 \subseteq I_1 \subseteq \cdots \subseteq I_n \subseteq \cdots$ is any ascending chain of right ideals of R , let I denote their union. If E_n denotes the injective hull of the difference module $R-I_n$, $n=0,1,\ldots,$ let E denote the direct sum of $\{E_n \mid n=0,1,\ldots\}$. The natural homomorphism $f_n\colon I \to R-I_n$ maps I into E_n . Furthermore, if $a \in I$, then $a \in I_k$ for some k , and then $f_n(a) = 0 \ \forall\, n \geq k$. Thus the symbol $\Sigma_{i=0}^{\infty} f_i(a) = f(a)$ is defined $\forall\, a \in I$, and $a \to f(a)$ is a map of I in E . Now $I \subseteq R \subseteq E_0$. Since E is quasi-injective by assumption, there exists $\lambda \in \mathrm{Hom}_R(E,E)$ which induces f . If $m = \lambda(1)$, where 1 is the identity element of R , then $\lambda x = mx \ \forall\, x \in R$. Clearly $m \in \Sigma_{i=1}^{t} E_i$ for some t, and then $f(I) \subseteq mR \subseteq \Sigma_{i=1}^{t} E_i$. It follows that $I_{t+1} = I_{t+2} = \cdots = I$, and R is therefore right noetherian.

Conversely, assume that R is right noetherian with identity, and let $M = \Sigma_{j\in J} \oplus M_j$ be a direct sum of a family $\{M_j \mid j \in J\}$ of injective modules. If I is any right ideal of R , then there exist $a_1, \ldots, a_n \in R$ such that $I = \Sigma_1^n a_i R$. If f is any map of I in M , then there is a finite subset P of J such that

$$f(a_i) \in N = \Sigma_{p \in P} M_p \ , \quad i=1, \ldots, n,$$

and then $f(I) \subseteq N$. Since N is injective, by Baer's criterion for unital modules there exists $m \in N$ such that $f(x) = mx \ \forall \ x \in I$. M therefore satisfies Baer's condition, and is accordingly injective.

The proof of the first part has the following consequence:

6. COROLLARY. If R is a ring, and if each direct sum of countably many injective modules is quasi-injective, then R is right noetherian.

The implication (A) $\rightarrow$ (B) is due to H. Bass [1], and the converse is stated as an exercise in Cartan-Eilenberg [1].

We now consider an important class of quasi-injective modules.

DEFINITION. A module M_R is semisimple in case M is a direct sum of simple submodules.

7. PROPOSITION. Let M_R be a module. Then M_R is semisimple if and only if M_R is a sum of simple submodules. In fact: if $M = \Sigma_{i \in I} M_i$, where $\{M_i \mid i \in I\}$ is a family of simple submodules, and if B is any submodule $\neq M$, then there exists a non-empty subset J of I such that the sum $M_J = \Sigma_{j \in J} \oplus M_j$ is direct, and such that $M = B \oplus M_J$.

PROOF. We prove the second assertion first, since the $B = 0$ case implies the first assertion.

If $B \cap M_i \neq 0$, then $B \cap M_i$ is a nonzero submodule of M_i, and simplicity of B implies that $B \supseteq M_i$. Since $B \neq M$, there exists $j \in I$ such that $B \cap M_j = 0$. For each subset $Q \subseteq I$ let $M_Q = \Sigma_{q \in Q} M_q$. Hence, the set S consisting of all subsets Q of I such that $B \cup \{M_q \mid q \in Q\}$ is an independent family of submodules is non-empty. If $\{Q_\lambda \mid \lambda \in \Lambda\}$ is a chain in S, and if $Q = \cup_{\lambda \in \Lambda} Q_\lambda$, then it is easy to see that the family $B \cup \{M_q \mid q \in Q\}$ is independent along with the families $B \cup \{M_q \mid q \in Q\} \ \forall \lambda$. Hence by Zorn's lemma S contains a maximal element, say Q. If $B + M_Q \neq M$, then there would exist $i \in I$ such that $(B+M_Q) \cap M_i = 0$, and then the family

$$B \cup \{M_q \mid q \in Q'\} \ , \quad Q' = Q \cup \{i\}$$

is independent, in violation of the maximality of Q . Hence, $M = B \oplus M_Q$, $M_Q = \Sigma_{q\in Q} \oplus M_q$, as asserted.

8. PROPOSITION. The following statements about a module M_R are equivalent:

(a) Each submodule of M is a sum of simple submodules.

(b) M is a sum of simple submodules.

(c) M is semisimple.

(d) Each submodule B of M is a direct summand of M: $= B \oplus K$, for some submodule K .

PROOF. (a) $\Rightarrow$ (b) is trivial, (b) $\Longleftrightarrow$ (c) is the first statement of 7, while (c) $\Rightarrow$ (d) follows from the second statement in 7, (d) $\Rightarrow$ (a). Assuming (d), we wish to prove that any submodule B is a sum of simple submodules, and we can assume that $B \neq 0$. If C is any submodule of B , then (d) implies that $M = C \oplus D$ for some submodule D of M , and then $B = C \oplus (B \cap D)$. This shows that B also satisfies property (d). If $0 \neq x \in B$, then Zorn's lemma leads to the existence of a submodule N of B which is maximal in the set of those submodules of B not containing x . Write $B = N \oplus P$, where P is a submodule of B . We assert that P is a simple module. For suppose T is a module satisfying $0 \subset T \subset P$. Then $P = T \oplus Q$, where Q is a submodule, and we assert that either $N \oplus T$ or $N \oplus Q$ violates the maximality of N . This can be seen as follows: if $x \in (N \oplus T) \cap (N \oplus Q)$, then write

$$x = x_1+t = x_2+q \qquad x_i \in N,\ i=1,2,\ t \in T,\ q \in Q .$$

Then $x_1-x_2 = q-t \in N \cap P$, but as $N \cap P = 0$, we conclude that $q = t \in Q \cap T$. Since $Q \cap T = 0$, then $q = t = 0$, so that $x = x_1 \in N$, a contradiction. This establishes that P is simple. Now write $B = C \oplus D$, where C is the sum of the simple submodules of B . Since D can contain no simple submodules, by what we have just proved it follows that $D = 0$, and then $B = C$. This proves (d) $\Rightarrow$ (a), and completes the proof of the proposition.

9. COROLLARY. Each semisimple module is quasi-injective.

PROOF. If M_R is semisimple, and if B is any submodule, then 8 implies that B is a direct summand of M. Consequently, any map of B into M can be extended to a map of M into M, proving that M is quasi-injective.

10. COROLLARY. A module M_R is semisimple if and only if M is the only essential submodule.

PROOF. Let M be semisimple, and let B be an essential submodule. Then B is a direct summand of M: $M = B \oplus K$, where K is a submodule. But $B \cap K = 0$ implies that $K = 0$, so $B = M$.

Conversely, let B be any submodule of M, and let K be a complement of B in M. Since $B+K$ is an essential submodule of M by 2.6, if M is the only essential submodule, then $M = B+K$. Since $B \cap K = 0$, actually $M = B \oplus K$, so M is semisimple by 8.

11. THEOREM. Let R be a ring. Then the following statements are equivalent.

(A) R is right artinian ring $\neq 0$, and R contains no nilpotent ideals $\neq 0$.

(B) R contains an identity element and R_R is semisimple.

(C) Each unital module M_R is injective.

(D) Each unital module M_R is semisimple.

PROOF. By the Wedderburn-Artin Theorem (3.21), if (A) holds, then R is a ring direct sum

$$R = R_1 \oplus \dots \oplus R_t ,$$

where each R_i is a f.d. full ring. Since each R_i has an identity, so does R. Since each right ideal of R_i is a right ideal of R, in order to show that R_R is semisimple, it suffices to show the same for each R_i. Hence assume that R is a f.d. full ring. Then (3.17) $R = D_n$, where D is a field. If $M = \{e_{ij} \mid i, j=1, \dots, n\}$ is a full set of matrix units in R, it is easily checked that $e_{ii}R$ is a minimal right ideal of R, $i=1, \dots, n$, and that $R = \Sigma_{i=1}^{n} \oplus e_{ii}R$. Thus R is semisimple, proving that (A) $\rightarrow$ (B).

(B) $\rightarrow$ (A). If R is semisimple, then the existence of the identity element in R implies that R is a direct sum of a finite number of simple right R-modules (= minimal right ideals). Then R_R is artinian (also noetherian) by 3. If N is any nilpotent right ideal of R, then, by 8, $R = N \oplus K$, where K is a right ideal of R. Writing $1 = e+f$, with $e \in N$, $f \in K$, it is easy to see that $e^2 = e$ and $f^2 = f$. Since e is necessarily nilpotent, we conclude that $e = 0$ and $1 = f \in K$, so that $K = R$, and then $N = 0$. This completes the proof of (B) $\Rightarrow$ (A).

(B) $\Rightarrow$ (C). If I is any right ideal of R, then I is a direct summand of R by 8. Consequently, each map f of I into a unital module M_R can be extended to a map g of R into M. Letting $m = g(1)$, it follows that $f(x) = mx \ \forall\, x \in I$; consequently M_R satisfies Baer's condition (§1). Then by Baer's criterion for unital modules, M_R is injective.

(C) $\Rightarrow$ (D). If M_R is unital, then each submodule is unital, hence injective. Thus (1.14), each submodule of M is a direct summand; consequently M is semisimple by 8. Since (B) is a special case of (D), the proof of 11 is complete.

REMARK. Recently Barbara Osofsky [1] proved that if R is a ring with identity such that every cyclic unital module over R is injective, then R is semisimple.

7. RATIONAL EXTENSIONS AND LATTICES OF CLOSED SUBMODULES

If $M_R \supseteq N_R$, then M is a <u>rational extension</u> of N in case for each submodule B, $M \supseteq B \supseteq N$, $f \in \mathrm{Hom}_R(B,M)$ satisfies $f(N) = 0$ if and only if $f = 0$. An equivalent property is: $\mathrm{Hom}_R(B-N,M) = 0$, for each such submodule B. We let $(M \blacktriangledown N)_R$ denote the condition that M is rational over N, and $(M \triangledown N)_R$ will signify that M is essential over N.

1. LEMMA. <u>If</u> $(M \blacktriangledown N)_R$ <u>then</u> $(M \triangledown N)_R$.

PROOF. Let F be a submodule of M such that $F \cap N = 0$. Then the sum $B = F + N$ is direct: $B = F \oplus N$. Now define $f \in \mathrm{Hom}_R(B,M)$ as follows: $f(x + y) = x$ for all $x + y \in B$, with $x \in F$, $y \in N$. Since $M \blacktriangledown N$, and since $f(N) = 0$, it follows that $f = 0$, so that $F = 0$. Thus, $M \triangledown N$.

2. PROPOSITION. $(M \blacktriangledown N)_R$ if and only if for each x, $0 \neq y \in M$, there exist $r \in R$, and $n \in Z$, such that $xr + xn \in N$, and $yr + yn \neq 0$.

PROOF. Sufficiency. Let B be a submodule such that $M \supseteq B \supseteq N$, let $f \in \mathrm{Hom}_R(B,M)$ be such that $f(N) = 0$, assume for the moment that $y = f(x) \neq 0$ for some $x \in B$, and choose r,n as in the statement such that $xr + xn \in N$, and $yr + yn \neq 0$. Then $0 = f(xr + xn) = yr + ny$, which is a contradiction. Thus, $f(N) = 0$ implies $f = 0$, and so $M \blacktriangledown N$.

Necessity. Assume that $M \blacktriangledown N$. let

$$x_N = \{(r,n) \in RXZ \mid xr + nx \in N\} ,$$

and let

$$y_0 = \{(r,n) \in RXZ \mid yr + ny = 0\} .$$

We must show that if $y_0 \supseteq x_N$, then $y = 0$. Let B denote the submodule of M generated by N and x. Then the typical element $b \in B$ has the form: $b = a + x(r,n)$ for some $a \in N$, $(r,n) \in RXZ$, where $u(r,n) = ur + nu \ \forall \ u \in M$. Consider the correspondence:

$$f: \ a + x(r,n) \to y(r,n).$$

If $a + x(r,n) = a' + x(r',n')$, with $a' \in N$, $(r',n') \in R \times Z$, then $x(r-r', n-n') = a' - a \in N$, so $(r-r', n-n') \in x_N$, and then $y(r-r', n-n') = 0$ by the assumption $y_0 \supseteq x_N$. Then $y(r,n) = y(r',n')$, and f is single-valued. Clearly $f \in \mathrm{Hom}_R(B,M)$, and obviously $f(N) = 0$. Since $M \blacktriangledown N$, this implies that $f = 0$. Then $y = f(x) = 0$, as asserted, and the proposition is proved.

3. COROLLARY. If $(M \blacktriangledown N)_R$, then for each submodule F of M, $(F \blacktriangledown F \cap N)_R$.

PROOF. Let $x, 0 \neq y \in F$. Then, by the proposition, there exist $r \in R$, $n \in Z$, such that $xr + nx \in N$, and $yr + ny \neq 0$. But $xr + nx \in F \cap N$, so by the proposition, $F \blacktriangledown (F \cap N)$.

4. PROPOSITION. If $M_R \supseteq N_R$, then (1) $Z(N) = Z(M) \cap N$, and if $M \triangledown N$, then $Z(M) \triangledown Z(N)$. (2) If $M \triangledown N$, then $Z(M) = 0$ if and only if $Z(N) = 0$.

PROOF. $Z(N) = Z(M) \cap N$ is obvious. If $M \triangledown N$, then $S \triangledown (S \cap N)$ for each submodule S of M. In particular, (1) holds. (2) is an immediate consequence of (1).

5. PROPOSITION. If $M_R \supseteq N_R$, and if $Z(N) = 0$, then $M \triangledown N$ if and only if $M \blacktriangledown N$.

PROOF. 1. supplies the "if" part. Now suppose $M \triangledown N$. Then $Z(M) = 0$ by 4. Let $x, 0 \neq y \in M$, and let $x_N = \{r \in R \mid xr \in N\}$, $y_0 = \{r \in R \mid yr = 0\}$. Since $Z(M) = 0$, and since $y \neq 0$, y_0 is not an essential right ideal of R, so $y_0 \not\supseteq x_N$. Hence there exists $r \in x_N$ such that $yr \neq 0$. Then $M \blacktriangledown N$ by 2.

DEFINITION. If $G_R \supseteq M_R$, and if $G \blacktriangledown M$, then G is a maximal rational extension of M in case G satisfies the following two conditions.

(1) If $F_R \supseteq M_R$, and if $F \blacktriangledown M$, then the identity map of M in M can be extended to a monomorphism of F_R into G_R.

(2) If $F_R \supseteq G_R$, and if $F \blacktriangledown M$, then $F = G$.

It can be shown that G_R satisfies (1) if and only if G_R satisfies (2).

6. THEOREM. Let $E = E(M_R)$, let $\Lambda = \mathrm{Hom}_R(E,E)$, and let $M^{\Lambda} = \{\lambda \in \Lambda \mid \lambda(M) = 0\}$. Then

$$\bar{M} = \bigcap_{\lambda \in M^{\Lambda}} \ker(\lambda)$$

is a maximal rational extension of M , and $\bar{M}$ contains each rational extension of M which is contained in E . If G_R is any maximal rational extension of M , then the identity map $M \to \bar{M}$ can be extended to a monomorphism of G onto $\bar{M}$.

PROOF. (a) Let B be any submodule of $\bar{M} \supseteq M$ (where $\bar{M}$ is defined as in the theorem), and let $f \in \mathrm{Hom}_R(B,\bar{M})$ be such that $f(M) = 0$. Since E_R is injective, f has an extension $f' \in \Lambda = \mathrm{Hom}_R(E,E)$. Since $f' \in M^{\Lambda}$, it follows that $\ker f' \supseteq \bar{M}$, so that $f'(B) = f(B) = 0$, that is, $f = 0$. Thus, $(\bar{M} \blacktriangledown M)_R$.

(b) Let F be any rational extension of $M \subseteq E$, let $t \in \Lambda$ be such that $t(M) =$ $= 0$, let

$$K = \{x \in F \mid t(x) \in M\} ,$$

and assume for the moment that $t(F) \neq 0$. Since $E \triangledown M$, it follows that $t(F) \cap M \neq 0$. Then, if $x \in F$ is such that $0 \neq y = t(x) \in t(F) \cap M$, then $x \in K$, so that $t(K) \neq 0$. Now K is a submodule of E , and t induces a mapping $t_0 \in \mathrm{Hom}_R(B,F)$, where $B = K + M$, and $F \supseteq B \supseteq M$.*) Since $F \blacktriangledown M$, and since $t_0(M) = 0$, it follows that $t_0 = 0$, that is, $t_0(K) = t(K) = 0$. Thus, the assumption $t(F) \neq 0$ leads to a contradiction. Hence, for each $t \in \Lambda$, $t(M) = 0$ implies $t(F) = 0$, so that $F \subseteq \bar{M}$. Thus, $\bar{M}$ is a rational extension of M , and $\bar{M}$ contains each rational extension of $M \subseteq E$.

(c) Now let H be a rational extension of M, $H \supseteq M$. Since E_R is injective, the identity map $M \to M$ can be extended to a homomorphism φ of H in E . Since $H \triangledown M$ by 1, φ is a monomorphism. Then $\bar{M} \supseteq \varphi(H)$ by (b) , so $\bar{M}$ satisfies (1) of the definition of a maximal rational extension.

(d) If $H \supseteq \bar{M}$, and if $H \blacktriangledown M$, then clearly $H \blacktriangledown \bar{M}$. Since $E = E(\bar{M})$, (c) shows that the identity map of $\bar{M} \to \bar{M}$ can be extended to a monomorphism φ of H in E. By (c) , $\varphi(H) \subseteq \bar{F}$ where $F = \bar{M}$. Clearly $\bar{F} = \bar{M}$, so $\varphi(H) \subseteq \bar{M} \subseteq \varphi(H)$, that is, $\varphi(H) = \bar{M}$. Thus, $\bar{M}$ satisfies (2) , so $\bar{M}$ is a maximal rational extension of M .

(e) Now let G be any maximal rational extension of M , and let $F = E(G)$. Then $F \cong E(M)$ under a map η which leaves fixed the elements of M . If $\bar{M}_F$ is defined for F as $\bar{M}$ was for E, it follows from (a) and (b) that $G = \bar{M}_F$. Then η

---

*) Note: $B = K$.

induces an isomorphism of G and $\bar{M}$ which extends the identity map $M \to \bar{M}$. This completes the proof of the theorem.

7. PROPOSITION. Let M_R be such that $Z(M_R) = 0$. Then each submodule N of M is contained in a unique closed essential extension N' contained in M.

PROOF. Let $E = E(M_R)$. Then 4 implies that $Z(E) = 0$; therefore $Z(N) = 0$ for each submodule N of E. From §2, we know that the closed submodules of E are the direct summands of E, so these modules are all injective. Let N be any submodule of E, and let F and G be two closed submodules of E which are essential extensions of N. Then F and G are injective hulls of N. If $x \in F$ (resp. $y \in G$) then $(N:x) = \{r \in R \mid xr \in N\}$ (resp. $(N:y)$) is an essential right ideal of R. Consequently, if $0 \neq u = x + y \in F + G$, $x \in F$, $y \in G$, then $I = (N:x) \cap (N:y)$ is an essential right ideal of R. Since $Z(E) = 0$, it follows that $uI \neq 0$. Since $uI \subseteq N$, this shows that $uR \cap N \neq 0$ for each $0 \neq u \in F + G$, and $F + G$ is therefore an essential extension of N. Then $(F + G) \nabla F$. Since F is injective, this implies that $F + G = F$, so that $G \subseteq F$. By symmetry $F \subseteq G$, so that $F = G$. This shows that N has a unique injective hull $F = N^*$ contained in E, completing the proof.

If $N \subseteq M$, then $N^* \cap M = N'$ must contain each essential extension of N contained in M. But $N^* \nabla N'$, so N^* is also the unique injective hull of N' contained in E. Thus, $(N')' = N'$, so N' must be closed. This proves the proposition.

Since $N \to N'$ is a closure operation, it is easy to see that arbitrary intersections of closed submodules of M_R is a closed submodule of M_R. It follows that the set $C(M_R)$ of closed submodules is a complete lattice with set intersection being the meet operation. If $\{C_i \mid i \in I\}$ is any collection of closed submodules of M, then $\bigvee_{i \in I} C_i = \left(\Sigma_{i \in I} C_i\right)'$.

8. COROLLARY. Let $Z(M_R) = 0$, and let P_R be an essential extension of M_R. Then $C(P_R)$ and $C(M_R)$ are complete lattices, and the contraction map $\varphi: C \to C \cap M$ is a lattice isomorphism of $C(P_R)$ and $C(M_R)$.

PROOF. We can assume $M \subseteq P \subseteq E = E(M_R)$. It suffices to prove the corollary for the case $P = E$. We have noted in the proof of 7, that if $C \in C(E)$, that is, if C

is an injective submodule of E, then $(C \cap M)^* = C$. This shows that φ is both onto and 1-1 (with $N \to N^*$ inducing the inverse of φ). Since φ is order preserving, we conclude that φ is a lattice isomorphism.

8. THEOREM. Let M_R be any injective module. Then the following conditions are equivalent:

(1) Each submodule N of M has a unique injective hull $E(N)$ contained in M.

(2) Each submodule N of M has a unique maximal essential extension $\hat{N}$ contained in M.

(3) If N_1, N_2 are injective submodules of M, then $N_1 \cap N_2$ is injective.

If M_R satisfies any of these equivalent conditions, then:

(4) The sum $N_1 + N_2$ of any two injective submodules N_1, N_2 of M is injective.

PROOF. By §2, (1) $\longleftrightarrow$ (2). Actually, we stated (2) above to emphasize that $E(N)$ is a unique subset of M, and not just unique in the sense of isomorphism (which we already know).

(1) $\to$ (3): $N_1 \cap N_2$ has an injective hull $E(N \cap M)$ in N_1 (resp. N_2), since N_1 (resp. N_2) is injective. By (1), $N_1 \cap N_2 \supseteq E(N_1 \cap N_2) \supseteq N_1 \cap N_2$, so $E(N_1 \cap N_2) = N_1 \cap N_2$ is injective. (3) $\to$ (1). If F_1, F_2 are two injective hulls of $N \subseteq M$, then $F_1 \cap F_2$ is injective, so $F_1 \cap F_2 = F_1 = F_2$.

(3) $\to$ (4). Since $N_1 \cap N_2$ is injective, $N_1 \cap N_2$ is a direct summand of N_2, so there exists a submodule N_2' such that $N_2 = (N_1 \cap M_2) \oplus N_2'$. Now $N_1 + N_2 = N_1 \oplus N_2'$, and N_1, N_2' are injective. This shows that $N_1 + N_2$ is injective.

By 7, we have the following:

9. COROLLARY. If M_R is injective, and if $Z(M_R) = 0$, then M_R satisfies the conditions of the theorem.

Eben Matlis [1] has proved the following interesting result concerning property (4).

10. PROPOSITION. The following two statements about a ring R are equivalent.

(A) Each epimorph of an injective module is injective.

(B) Each sum $M_1 + M_2$ of injective submodule of a module M_R is injective.

PROOF. Let first E be any module, and N a submodule. Let $Q = (E,E)$ be the direct product (= sum) of two copies of E, and let K denote the diagonal of (N,N), that is, let $K = \{(x,x) \in Q \mid x \in N\}$. In the difference module $\bar{Q} = Q-K$, let $M_1 =$ $= \{y+K \in \bar{Q} \mid y \in (E,0)\}$, $M_2 = \{y+K \in \bar{Q} \mid y \in (0,E)\}$.

Then $\bar{Q} = M_1+M_2$. Since $(E,0) \cap K = o$ (resp. $(0,E) \cap K = 0$), clearly $M_i \cong E$, $i=1, 2$. Furthermore, $M_1 \cap M_2 = \{y+K \in \bar{Q} \mid y \in (N,0)\} = \{y+K \in \bar{Q} \mid y \in (0,N)\}$, and $M_1 \cap M_2 \cong N$ under $y \to y+K \ \forall \ y \in (N,0)$.

(B) $\to$ (A). If E is an injective module, then M_1, M_2 are injective, hence so is $\bar{Q} = M_1+M_2$. Since M_1 is injective, there exists a submodule G of $\bar{Q}$ such that $\bar{Q} = M_1 \oplus G$, and G is necessarily injective. But

$$G \cong (M_1+M_2) - M_1 \cong M_2 - (M_1 \cap M_2).$$

Since $M_2 \cong E$, and since $M_1 \cap M_2 \cong N$, it follows that $E-N$ ($\cong G$) is injective. Therefore each epimorph $E+N$ of an injective module E is injective.

(A) $\to$ (B). Let M_1, M_2 be injective submodules of M. Then $M_1 + M_2$ is an epimorph of the direct sum $M_1 \oplus M_2$. Since $M_1 \oplus M_2$ is injective, it follows that $M_1 + M_2$ is injective.

Rings with the property 10 are called Left hereditary rings (Cartan-Eilenberg [1]).

Barbara Osofsky has noted the following:

11. COROLLARY. Let R be a ring with identity such that each unital module M_R has the equivalent properties (1)-(3) of 8. Then R is semisimple artinian.

PROOF. Let N_R be any unital module, and let $E = E(N_R)$. As in the proof of 10, N is the intersection $M_1 \cap M_2$ of two injective submodules of $(E \oplus E) - K$. Thus, each unital module N_R is injective, so R_R is semisimple artinian by §6.

8. MAXIMAL QUOTIENT RINGS

A ring A containing a ring R is a right quotient ring of R in case A_R is a rational extension of R_R, notationally $(A \blacktriangledown R)_R$. Quotient rings in this generality were first studied by Utumi [1]. In case R has right singular ideal $Z_r(R) = 0$, these coincide with the quotient rings of R.E. Johnson [3,4].

A. LEMMA. If A and B are right quotient rings of R, and if $\varphi: A_R \to B_R$ is module monomorphism of A_R into B_R which induces the identity map on R, then φ is ring monomorphism of A into B.

PROOF. If x, y are elements of A (resp. of B), let $x \circ y$ (resp. xy) denote the product in A (resp. in B). Thus, $\varphi(a \circ r) = \varphi(a)r$, whenever $a \in A$ and $r \in R$. For notational simplicity, consider $A \subseteq B$, that is, identify a with $\varphi(a)$. Then $a \circ r = ar$ $\forall\, a \in A,\ r \in R$. If $a \in A$, the correspondence $f_1: x \to a \circ x$, defined $\forall\, x \in A$, is a map of the module A_R into A_R. Similarly, $f_2: x \to ax$ is a map of A_R into B_R (conceivably $aA \not\subseteq A$). Then $f = f_1 - f_2$ is a map of A_R into B_R and $f = 0$ on R. Since B_R is a rational extension of R_R, necessarily $f = 0$ on A, that is, $a \circ x = ax$ $\forall\, x \in A$. Since this is true for all $a \in A$, this shows that ring multiplication in A is induced by that of B. In other words, $\varphi(a \circ b) = \varphi(a)\varphi(b)$ $\forall\, a, b \in A$, so φ is a ring monomorphism as stated.

B. THEOREM. Let R be a ring with identity, E the injective hull of R_R, $\bar{R}$ the maximal rational extension of R in E. Then:

(1) $\bar{R}$ is a ring whose operations $\bar{R} \times \bar{R} \to \bar{R}$ induce the module operations $\bar{R} \times R \to \bar{R}$ in $\bar{R}_R$; in particular R is a subring of $\bar{R}$.

(2) If A is a right quotient ring of R, then there is a ring monomorphism φ of A into $\bar{R}$ which induces the identity map on R.

PROOF. Let $\Lambda = \mathrm{Hom}_R(E,E)$, and let $Q = \mathrm{Hom}_\Lambda(E,E)$. The elements of Λ (resp. Q) are written on the left (resp. right) and E is a (Λ,Q)-bimodule. Thus, if $\lambda \in \Lambda$, $x \in E$, $q \in Q$, then $(\lambda x)q = \lambda(xq)$. By 7.6, $\bar{R}$ is the double annihilator of R in Λ, that is, if $R^\wedge = \{\lambda \in \Lambda \mid \lambda R = 0\}$, then $\bar{R} = \{m \in E \mid R^\wedge m = 0\}$. Since R has an identity element 1, $R^{\Lambda} = \{\lambda \in \Lambda \mid \lambda 1 = 0\}$.

Now Q contains a subring R' which is ring homomorphic to R under the mapping $r \to r'$, where $r' \in Q$ is the map $x \to xr$ defined for all $x \in E$. Since Q is therefore a natural right R-module, and since Q is a ring whose "multiplication" $Q \times Q \to Q$ induces the right R-module operation $Q \times R \to Q$ of Q_R, in order to prove (1), it suffices to show that the right R-modules Q_R and $\bar{R}_R$ are isomorphic. If $q \in Q$, then the element $1q = q^o \in E$, and

$$(qr)^o = (qr')^o = 1(qr') = (1q)r' = q^o r' = q^o r$$

$\forall\, r \in R$. Thus, the correspondence $q \to q^o$ is a homomorphism θ of Q_R into E_R. We shall show that θ is an isomorphism of Q_R and $\bar{R}_R$.

(a) θ is a monomorphism. Let $x \in E$. Then the correspondence $r \to xr$, defined $\forall\, r \in R$, is a map of R_R into E_R which by the injectivity of E_R is induced by an element $\lambda \in \Lambda$. Then $\lambda(1) = x$. If $q \in Q$ is such that $q^o = 0$, then

$$xq = (\lambda \cdot 1)q = \lambda(1q) = \lambda(q^o) = \lambda(0) = 0 .$$

Since this is so $\forall\, x \in E$, necessarily $q = 0$, so θ is a monomorphism.

(b) θ maps Q into $\bar{R}$. If $q \in Q$, and if $\lambda \in R^{\Lambda}$, that is, if $\lambda(1) = 0$, then

$$\lambda(q^o) = \lambda(1(q)) = \lambda(1)q = 0 \cdot q = 0 ,$$

proving that $q^o \in \bar{R}$, $\forall\, q \in Q$. Thus, $Q^o = \theta(Q) \subseteq \bar{R}$.

(c) θ maps Q onto $\bar{R}$. If $m \in E$, there exists $\lambda_m \in \Lambda$ such that $\lambda_m(1) = m$. Let x be a fixed element of $\bar{R}$, and consider a correspondence $k\colon m \to \lambda_m(x)$ defined $\forall\, m \in M$. If $\lambda' \in \Lambda$ is such that $\lambda'(1) = m$, then $(\lambda' - \lambda_m)(1) = 0$, so $\lambda' - \lambda_m\ R^{\Lambda}$. Then $(\lambda' - \lambda_m)\bar{R} = 0$, and $\lambda'(x) = \lambda_m(x)$. Hence, the correspondence $m \to \lambda_m(m)$ is independent of the choice of λ_m, and is therefore a mapping of E into E. If $\mu \in \Lambda$, then

$$(\mu m)k = \lambda_{\mu m}(x) = \mu\lambda_m(x) = \mu((m)k) ,$$

showing that $k \in Q$. But $(1)k = \lambda_1(x) = x$, showing that $Q^o = \bar{R}$. This completes the proof of (1) of the theorem.

(2) If A is a quotient ring of R, then by 7.6, there is a monomorphism φ of the module A_R into $\bar{R}_R$ which induces the identity map on R, and φ is a ring monomorphism by Lemma A.

All quotient rings will be right quotient rings unless stated contrarywise. A quotient ring A of R is <u>maximal</u> in case given any quotient ring T of R there exists a ring monomorphism of T into A which induces the identity map on R. The theorem shows that if R is a ring with identity, then $\bar{R}$ is a maximal quotient ring of R. The material below up to and including Theorem 1 is largely devoted to handling the case of the missing identity.

C. PROPOSITION. If R is a ring, then any two maximal quotient rings of R are isomorphic under a mapping which induces the identity mapping of R.

PROOF. If A,B are maximal right quotient rings of R, then there exist ring monomorphisms $\varphi_1: A \to B$ and $\varphi_2: B \to A$ which extend the identity mapping of R. If 1_A denotes the identity mapping of A, then $\varphi_2\varphi_1 - 1_A = 0$ on R. Since $(A \blacktriangledown R)_R$, this implies that $\varphi_2\varphi_1 - 1_A = 0$ on A, that is, $\varphi_2\varphi_1 = 1_A$. By symmetry, $\varphi_1\varphi_2 = 1_B$, so φ_1 and φ_2 are inverses of each other, and $\varphi_1: A \to B$ is a ring isomorphism.

D. PROPOSITION. (1) If $A \supseteq B \supseteq R$ are rings, and if A is a quotient ring of B and B is a quotient ring of R, then A is a quotient ring of R. (2) If B is a maximal quotient ring of R, then B is its own maximal quotient ring.

PROOF. (1) By 7.2, we must show that given $x, y \in A$, and $y \neq 0$, that there exists r_1 in the ring $R_1 = R+Z$ obtained from R by freely adjoining an identity element ($R_1 = R$ if R has an identity) such that $xr_1 \in R$ and $yr_1 \neq 0$. Since $(A \blacktriangledown B)_B$, by 7.2, there exists b_1 in the ring B_1 (defined for B similarly as R_1 was defined for R) such that $xb_1 \in B$ and $yb_1 \neq 0$. By the same reasoning, there exists $b_2 \in B_1$ such that $yb_1b_2 \in B$ and $yb_1b_2 \neq 0$. Thus, $b' = b_1b_2 \in B$ is such that $xb', yb' \in B$ and $yb' \neq 0$. Now $b' = b+n$, with $b \in B$ and $n \in Z$. Since $(B \blacktriangledown R)_R$, there exists $r' \in R$, such that $br' = r \in R$, and $yb'r' \neq 0$. Then $r'' = b'r'$ lies in R_1 and $yr'' \neq 0$. Since $yb' \in B$, $yb'r' = yr'' \in B$, so we can choose $r''' \in R$, such that

$xr''r''' \in R$ and $yr''r''' \neq 0$. Then $r_1 = r''r''' \in R_1$, is such that $xr_1 \in R$ and $yr_1 \neq 0$, as desired. (No doubt there is a more satisfying way of doing this!)

(2) If B is a maximal quotient ring of R, and if A is a quotient ring of B, then A is a quotient ring of R by (1). But B is a maximal rational extension of R. Hence, by the definition of maximal rational extensions preceding 7.6, $A = B$.

Let R be any ring. If $x \in R$, then x induces an endomorphism x^* of R_R: $x^*(a) = xa$, $\forall\, a \in R$. Furthermore, $(x+y)^* = x^*+y^*$, $(xy)^* = x^*y^*$, $\forall\, x, y \in R$, so $x \to x^*$ is a ring homomorphism of R into the ring $S = \mathrm{Hom}_R(R,R)$. The ring R is said to be <u>left faithful</u> in case this homomorphism is an isomorphism. (This is equivalent to the condition that $xR = 0$ for any $x \in R$ implies $x = 0$). Let R^* be the subring of S consisting of $\{r^* \mid r \in R\}$.

E. LEMMA. Let R be a left faithful ring, and let R' denote the subring of $S = \mathrm{Hom}_R(R,R)$ generated by R^* and the identity element e of S. (1) Let T be any ring with identity f containing R, and let T_o be the subring of T generated by R and f. Then there is a ring epimorphism φ of T_o on R'. (2) Moreover, φ is an isomorphism if and only if T_o is a quotient ring of R.

PROOF. (1) If $t \in T_o$, then $tR \subseteq R$, so t induces an element t' of S: $t'(a) = ta$ $\forall\, a \in R$. Clearly, the correspondence $\varphi: t \to t'$, defined for all $t \in T_o$, is a ring homomorphism of T_o into S. Furthermore $\varphi(r) = r^*$ $\forall\, r \in R$, and $\varphi(f) = e$, so that $\varphi(T_o) \supseteq R'$. If $t \in T_o$, then $t = r + nf$, with $r \in R$, $n \in Z$, and $\varphi(t) = r^* + ne \in R'$, showing that $\varphi(T_o) \subseteq R'$. Thus φ is the desired ring epimorphism.

(2) If $s \in R'$, and if $s \neq 0$, then $sR \neq 0$. But $sR \subseteq R$. Hence given $t, s \in R'$, and $s \neq 0$, there exists $r \in R$ such that $tr \in R$ and $sr \neq 0$. By 7.5, R' is a rational extension of R_R, that is, R' is a quotient ring of R. Thus, if φ is an isomorphism, T_o is a quotient ring of R. Conversely, if T_o is a quotient ring of R, and if $t \in T_o$, then t induces an endomorphism t_o of the right R-module T_o. Furthermore, the restriction of t_o to R is the endomorphism $t' = \varphi(t)$ of R_R. Thus, if $tR = 0$, then $t' = 0$, and the fact that T_o is a rational extension of R_R implies that $t_o = 0$. Then $t = t_o \cdot f = tf = t = 0$, so φ is an isomorphism.

Because of (1) of the lemma, we call R' the <u>unit-cover</u> of R.

F. PROPOSITION. Let R be any left faithful ring, R' its unit-cover, and let $\bar{R}$ denote the maximal rational extension of R_R. (1) $\bar{R}$ is a natural right R'-module, and the module operations $\bar{R} \times R' \to \bar{R}$ in $\bar{R}_{R'}$ induce the module operations $\bar{R} \times R \to \bar{R}$ in $\bar{R}_R$. (2) $\bar{R}$ is a maximal quotient ring of R. Any maximal quotient ring of R contains an identity element, and is isomorphic to $\bar{R}$ under a mapping which extends the identity mapping of R.

PROOF. (1) Let $t = r^* - ne$, $r \in R$, $n \in Z$, be any element of R'. First suppose that $t = 0$. Then $ra = na$ $\forall\, a \in R$. Let x be a nonzero element of $\bar{R}$. By 7.2, there exist elements $s \in R$, $m \in Z$ such that $xs - mx \in R$, and $xs - mx \neq 0$. Thus, if $xR = 0$, then $mx \neq 0$ and $mx \in R$ for some $m \in Z$. Suppose $y \in \bar{R}$ is such that $x = yr - ny \neq 0$. Since $xR = y(tR) = 0$, then $mx \neq 0$ and $b = mx \in R$ for some $m \in Z$. But $bR = mxR = 0$, and since R is left faithful, this implies that $b = 0$, a contradiction. Hence $t = 0$ implies that $yr - ny = 0$ $\forall\, y \in \bar{R}$. This means that given $t \in R'$, $y \in R'$, if $t = r^* - - ne = s^* - me$, for $r, s \in R$, $n, m \in Z$, then $yr - ny = ys - me$. Thus, the correspondence $(y, r^* - ne) \to yr - ne$ is a mapping of $\bar{R} \times R'$ into R'. Clearly, then $\bar{R}$ is a right R'-module whose operations induce those of $\bar{R}_R$.

(2) Since R' is a rational extension of R_R, we may assume that the module R'_R is embedded in $\bar{R}_R$. By (1), the operations of $\bar{R}_R$, induce the ring operations of R'. Since $\bar{R}$ is a rational extension of R, and since the module operations of $\bar{R}_R$, induce those of $\bar{R}_R$, $\bar{R}$ is a rational extension of $R'_{R'}$. Now let $A_{R'}$ be any rational extension of $R'_{R'}$. Then A is a right R-module, and A_R remains a rational extension of R'_R. Hence there exists a monomorphism $\varphi: A_R \to \bar{R}_R$, and clearly φ is a monomorphism of $A_{R'}$ into $\bar{R}_{R'}$. Thus, $\bar{R}$ is a maximal quotient ring of R'. Thus, $\bar{R}$ is a quotient ring of R.

Now let A be any quotient ring of R. The proof of (2) of Theorem B did not use the fact that the ring R had an identity. Thus, A can be embedded in $\bar{R}$ by a ring monomorphism φ which induces the identity map of R. The uniqueness of maximal quotient rings of R follows from Proposition C. Thus, any maximal quotient ring of R has an identity, since $\bar{R}$ has. This completes the proof.

The next lemma could have been proved in §7, but we postponed it for use in the proof of the theorem following.

G. LEMMA. If M_R is an injective module with singular submodule $Z(M) = 0$, then the endomorphism ring Λ of M_R is a right self-injective von Neumann (regular) ring.

PROOF. Since $Z(M) = 0$, then $Z(N) = 0$ for any submodule N. Thus, if M is an essential extension of N, $M \triangledown N$, then M is a rational extension of N, $M \blacktriangledown N$, by 7.5. Let J denote the radical of Λ. If $\lambda \in J$, then $M \triangledown (\ker \lambda)$. Since $\lambda(\ker \lambda) = 0$, and since $M \blacktriangledown (\ker \lambda)$, then $\lambda = 0$. Then $J = 0$, and the lemma is a direct consequence of (2) of 5.1.

1. + 2. THEOREM. Let R be a ring with right singular ideal $Z_r(R) = 0$, let E denote the injective hull of R_R, and let $\Lambda = \mathrm{Hom}_R(E,E)$. Then: (1) Λ is a right R-module isomorphic to E_R; (2) E is a ring, a maximal quotient ring of R, and the isomorphism $\varphi\colon \Lambda_R \to E_R$ in (1) is a ring isomorphism of Λ and E; (3) Λ (resp. E) is a right self-injective von Neumann (regular) ring.

PROOF. By Lemma G, Λ is right self-injective and von Neumann (regular). Hence it remains only to prove (1) and (2). Since $Z_r(R) = 0$, E is a rational extension of R_R by 7.5; E is therefore the maximal rational extension of R_R. By Proposition F, E is a ring with identity, and is the maximal quotient ring of R.

If $r \in R$, then r induces a endomorphism r^* of R_R, where $r^*(x) = rx$ $\forall\, x \in R$. Since E_R is injective r^* is induced by an element of Λ. Since E is a rational extension of R, r^* is induced by a unique element of Λ, which we also denote by r^*. Moreover, the mapping $r \to r^*$ is ring isomorphism of R into Λ. Thus, Λ is a natural right R-module, under the operation $\lambda r = \lambda r^*$ defined for $\forall\, \lambda \in \Lambda$, $r \in R$. For simplicity, identify r with r^*.

Let R' denote the subring of E generated by R and the identity element 1 of E. Then E is a natural right R'-module, E is the injective hull of R' considered as a right R'-module, $\Lambda = \mathrm{Hom}_{R'}(E,E)$, and E is a maximal quotient ring of R'. All this follows from Proposition F, or its proof.

As in the proof of Theorem B, if $x \in E$, there is an element $\lambda_x \in \Lambda$ such that $\lambda_x(1) = x$. Since E is a rational extension of R', the element λ_x is uniquely determined by x. Clearly $\lambda_{x+y} = \lambda_x + \lambda_y$, and $\lambda_{xr} = \lambda_x r$, $\forall\, x, y \in E$, $r \in R$, so

$\theta: x \to \lambda_x$ is a monomorphism of E_R into Λ_R. If $\lambda \in \Lambda$, then $\lambda = \lambda_y$, where $y = \lambda(1)$, so θ is an isomorphism of E_R and Λ_R. This means that Λ_R is a rational extension of R_R (along with E_R), so Λ is maximal quotient ring of R. By Lemma A, θ is a ring isomorphism of E and Λ. This completes the proof.

DEFINITION. <u>At times the term quotient ring will be abbreviated to</u> q.r., <u>and maximal quotient ring to</u> m.q.r.

3. PROPOSITION. A ring R has a q.r. which is a von Neumann*) ring if and only if $Z_r(R) = 0$. In particular: if R is a von Neumann ring, then $Z_r(R) = 0$.

PROOF. If $Z_r(R) = 0$, then E of 1 is a q.r. and a von Neumann ring. Conversely, let Q be a q.r. of R which is a von Neumann ring. For each $b \in Q$, $I_b = \{r \in R \mid br = 0\}$ is a right ideal of R. If $a \in R$, $a \neq 0$, there exists $x \in Q$ such that $a \times a = a$. Then $e = xa$ is an idempotent $\neq 0$, and $I_e = I_a$. Now $P = \{y - ey \in Q \mid y \in Q\}$ is a right ideal of Q, and $I_a = P \cap R$. Now $S = eQ \cap R \neq 0$ (since $Q_R \text{ '}\supseteq R_R$), and also $S \cap I_a = 0$. This shows that I_a is not an essential right ideal of R when $0 \neq a \in R$. Thus, $Z_r(R) = 0$.

The last remark is an immediate consequence, since R is a q.r. of R.

For any ring T we let $C_r(T)$ denote the totality of closed submodules of T, that is, $C_r(T) = C(T_T)$.

4. THEOREM. Let R be a ring with $Z_r(R) = 0$ and let A be any quotient ring of R. Then: (1) $C_r(A) = C(A_R)$; (2) $C_r(A)$ is isomorphic to $C_r(R)$ under contradiction; (3) if S is the maximal quotient ring of R, then $C_r(S)$ consists of the principal right ideals of S (= the set of ideals of S generated by idempotents.)

PROOF. By 2, we can assume $S \supseteq A \supseteq R$, where S is the m.q.r. (= maximal q.r.) of R. First assume $A = S$. Then, by 1, S_R is the injective hull of R_R. By §7, we have: (i) $C(S_R)$ consists of direct summands of S_R; (ii) $C(S_R) \cong C(R_R)$ under contraction; and (iii) if $I \in C(R_R)$, the corresponding $I^* \in C(S_R)$ is the injective hull

*) Called a regular ring in the literature. See the Introduction.

of I_R . Since $Z(R_R) = 0$, and since $(S \nabla R)_R$, it follows from §7, that $(S \blacktriangledown R)_R$.

We first show that each $I^* \in C(S_R)$ is a right ideal of S . Let $a \in I^*$, and let $x = b + as \in I^* + aS$, where $b \in I^*$, $s \in S$. Since $(S \blacktriangledown R)_R$, by 7.2, if $0 \neq y \in I^* + aS$, there exist $r \in R$, $n \in Z$ such that $sr' \in R$, and $yr' \neq 0$, where $r' = r + n$. Then

$$xr' = br' + a(sr') \in I^* \quad \text{and} \quad yr' \neq 0 .$$

By 7.2, this shows that $(I^* + aS)_R$ is a rational, hence essential, extension of I^*_R . Since I^*_R is injective, we conclude that $I^* = I^* + aS$, so that $aS \subseteq I^*$ $\forall$ $a \in I^*$. Thus, I^* is a right ideal of S .

Since $C(S_R) \cong C(A_R)$ under contraction, it follows that each $C \in C(A_R)$ is a right ideal of A . We next show that $C(A_R) = C(A_A) = C_r(A)$. By the arguments of 2.4, if $C \in C(A_R)$, then C is the complement in A_R of some $K \in C(A_R)$. Since C,K are right ideals of A , it follows that C (also $K) \in C(A_A) = C_r(A)$. Similarly, if $C \in C_r(A)$, then C is the complement in A_A of some $K \in C_r(A)$. Since C,K are R-submodules of A_R , there exists $Q \in C(A_R)$ such that $Q \supseteq C$ and such that Q is the complement in A_R of K . But Q is then a right ideal of A , so $Q = C$, and therefore $C \in C(A_R)$. This establishes that $C(A_R) = C_r(A)$. Since $C(A_R) \cong C(R_R)$ under contraction, this proof of (2) is completed by the substitutions $C_r(A) = C(A_R)$, $C_r(R) = C(R_R)$.

(3) By (i), $C(E_R)$ consists of the direct summands of E_R . It is easy to see that a right ideal of a ring with identity is a direct summand of that ring if and only if it is a principal right ideal generated by an idempotent. In a von Neumann ring, as is S , each principal right ideal is generated by an idempotent. This establishes (3).

For later use, we now establish a result for a ring with zero (right) singular ideal.

5. PROPOSITION. Let R be such that $Z_r(R) = 0$, and if H is a right ideal of R , let H' denote the smallest closed right ideal of R containing H. Then: (1) $aH' \subseteq (aH)'$ $\forall$ $a \in R$; (2) $(I:a) \in C_r(R)$ $\forall\, a \in R$ $\forall\, I \in C_r(R)$; (3) $C_r(R)$ contains the set $A_r(R)$ of annihilator right ideals of R .

PROOF. We shall use the fact that H' is the unique maximal essential extension of H in R .

(1) The result is trivial if $aH' = 0$. If $aH' \neq 0$, let ay be an arbitrary nonzero element of aH', $y \in H'$. Since $H' \nabla H$, $J = \{r \in R \mid yr \in H\}$ is an essential right ideal of R. Since $Z_r(R) = 0$, and since $ay \neq 0$, then $ayJ \neq 0$. Clearly, $ayJ \subseteq aH \cap [ay]$, where $[ay]$ denotes the submodule of aH' generated by ay. Thus, $(aH') \nabla (aH)$, and consequently $aH' \subseteq (aH)'$.

(2) By (1) we have

$$a(I:a)' \subseteq (a(I:a))' \subseteq I' = I ,$$

so that $(I:a)' \subseteq (I:a)$. This proves (2).

(3) Let $A = X^r$ be any annihilator right ideal, where $X \subseteq R$. Clearly $A = \cap_{a \in X}(0:a)$. Since $C_r(R)$ is a complete lattice, and since $(0:a) \in C_r(R)$ $\forall a \in X$ by (2), we conclude that $A \in C_r(R)$.

A ring R is (right, left) prime in case $R \neq 0$ and in case $IJ = 0$ for (right, left) ideals I,J of R implies that $I = 0$ or $J = 0$.

6. **PROPOSITION.** The following statements for a ring $R \neq 0$ are equivalent:

(a) R is right prime
(b) R is left prime
(c) R is prime
(d) $aRb = 0 \rightarrow a = 0$ or $b = 0$ $\forall$ $a,b \in R$.

PROOF. (a) → (c), (b) → (c) are trivial. Now assume (c), and let I,J be right ideals such that $IJ = 0$. Then (RI), (RJ) are ideals of R and $(RI)(RJ) = R(IR)J \subseteq RIJ = R0 = 0$. Hence either $RI = 0$ or $RJ = 0$. If $RI = 0$, then I is contained in $R^r = \{r \in R \mid Rr = 0\}$, which is an ideal of R satisfying $(R^r)^2 = 0$. By (c), $R^r = 0$, so $I = 0$. Thus (c) → (a). Similarly (c) → (b).

Assume (a), and let $aRb = 0$, with $a,b \in R$. Then $(aR)(bR) = 0$, so $aR = 0$ or $bR = 0$ by (a). If $aR = 0$, then $a \in R^l = \{x \in R \mid xR = 0\}$, which is an ideal of R satisfying $(R^l)^2 = 0$. By (a) $R^l = 0$, so $a = 0$. Thus, (a) → (d).

Assume (d), and let $IJ = 0$, where I,J are right ideals. Then $IRJ \subseteq IJ = 0$. If $I \neq 0$, pick $0 \neq a \in I$. Then $aRJ = 0$, so $J = 0$.

7. COROLLARY. If R is a prime ring, and if V is a nonzero left ideal of R, then R is a right quotient ring of any subring T which contains V.

PROOF. If $0 \neq a \in R$, then $aV \neq 0$ by the fact that R is (left) prime. But $aV \subseteq V$, so that $aT \cap T \neq 0$. Thus $R_R \ '\supseteq T_T$ as asserted.

The most common examples of prime rings are integral domains and primitive rings.

We now describe $\hat{R}$ in a special case.

8. THEOREM. Let R be a prime ring satisfying $Z_r(R) = 0$, and containing a minimal closed right ideal U. Then $\hat{R}$ is isomorphic to the full ring of l.t.'s in a right vector space over a field.

PROOF. By the lattice isomorphism $\varphi: C_r(\hat{R}) \to C_r(R)$, and by the fact that $\hat{R}$ is a von Neumann ring, it follows that $\varphi^{-1}U = W$ is a minimal right ideal of $\hat{R}$. Furthermore, $W = e\hat{R}$, $e = e^2 \in \hat{R}$, and then $V = \hat{R}e$ is a minimal left ideal of $\hat{R}$. Let $S = \hat{R}$, and let $L = \mathrm{Hom}_{eSe}(V,V)$. Then correspondence $r \to \bar{r}$ is a ring monomorphism of S in L, where $\bar{r}(x) = rx \quad \forall \quad x \in V,\ r \in S$.

(i) Since V is a minimal left ideal of S, $eSe \cong \mathrm{Hom}_S(V,V)$ is a field. Thus, L is a full ring of l.t.'s in the right vector space V over eSe.

(ii) $\bar{V} = \{\bar{v} \in L \mid v \in V\}$ is a left ideal of L.

To see this, let $a \in L$, and let $\bar{v} \in \bar{V}$. Then if $x \in V$,

$$a\bar{v}(x) = a(\bar{v}(x)) = a(v \cdot x)$$

also

$$\begin{aligned} \overline{a(v)}(x) &= a(v) \cdot x = a(ve)xe = a(v) \cdot exe \\ &= a(v \circ exe) = a(vx). \end{aligned}$$

Thus $a\bar{v} = \overline{a(v)} \in \bar{V} \quad \forall \quad a \in L,\ \bar{v} \in \bar{V}$, so $\bar{V}$ is a left ideal of L.

(iii) L is a right quotient ring of $\bar{S} = \{\bar{r} \in L \mid r \in S\}$. This follows from 7, since $\bar{S}$ contains $\bar{V}$ and $\bar{V}$ is a left ideal of L by (ii).

(iv) $L = \bar{S}$.

Since S_S is injective by 1, (iii) implies (iv). This completes the proof of the theorem, since $S \cong L$ by (iv), and L is a full linear ring by (i).

A ring R is semiprime if $R \neq 0$, and if R contains no nilpotent ideals $\neq 0$.

9. PROPOSITION. A ring $R \neq 0$ is semiprime if and only if R contains no nilpotent left (resp. right) ideals.

PROOF. The "if" part is trivial. Conversely let I be a nilpotent left ideal of R, and suppose R is semiprime. If $I^k = 0$, then $(IR)^{k+1} = I(RI)^k R \subseteq I(I)^k R = 0$, so IR is a nilpotent ideal. Then semiprimeness of R implies $IR = 0$. Then $I \subseteq R^1 = \{x \in R \mid xR = 0\}$. But R^1 is a nilpotent ideal of R, so $R^1 = 0$, and $I = 0$. The parenthetical part follows by symmetry.

We complete this section with two propositions which are preparatory to the next section.

10. PROPOSITION. If R has zero singular right ideal, and if R is semiprime, then R is prime if and only if $\hat{R}$ is prime.

PROOF. Suppose that R is prime, and let I,J be nonzero ideals of $\hat{R}$. Then $I \cap R$, $J \cap R$ are nonzero ideals of R. Since R is prime, $(I \cap R)(J \cap R) \neq 0$, and then $IJ \neq 0$, so $\hat{R}$ is prime.

Conversely let I,J be right ideals in R such that $JI = 0$. Then ISJ is a right R-submodule of $S = \hat{R}$, and $R \cap ISJ$ is a right ideal of R. Clearly $(R \cap ISJ)^2 = 0$, and so $R \cap ISJ = 0$, by semiprimeness of R. Since $S_R \,'\supseteq R_R$, it follows that $ISJ = 0$. If S is prime, then $J \neq 0 \to IS = 0$, and then $I = 0$. Thus, R is prime.

EXAMPLE. Let T denote the subring of lower triangular matrices of the full ring $S = D_n$ of $n \times n$ matrices over the field D. If $\{e_{ij} \mid i,j = 1,\dots,n\}$ are matrix units of S, then T contains the left ideal Se_{11} of S. If R denotes either T or S, by 7, we know that S is a right quotient ring of R. Then $Z_r(R) = 0$ by 3. Since S_S is injective, it follows* that $S = \hat{R}$. Furthermore, R contains a nilpotent left ideal, namely $I = (1 - e_{11})Re_{11}$. It is easy to check that I is a left ideal of

(*) Since $S \subseteq \hat{R}$, injectivity of S_S implies $\hat{R} = S \oplus K$, for some S-submodule K of R. But then K is an R-submodule of $\hat{R}$, and $K \cap R = 0$, so $K=0$ and $\hat{R}=S$ as asserted.

T and, hence, of R ; trivially $I^2 = 0$. This example shows that R need not be semiprime even though $Z_r(R) = 0$ and $\hat{R}$ is simple artinian. Furthermore, $\hat{R}$ is prime while R is not even semiprime.

11. **LEMMA.** Let R satisfy $Z_r(R) = 0$, and let e be a central idempotent of $\hat{R}$. Then $K = eR$ satisfies $Z_r(K) = 0$, and $e\hat{R} = \hat{K}$. If R is semiprime so is eR .

PROOF. $R \cap e\hat{R}$ is an essential R-submodule of $e\hat{R}$. Clearly $R \cap e\hat{R} \subseteq eR$, so $e\hat{R}$ is a right quotient ring of eR . Since $e\hat{R}$, $(1 - e)\hat{R}$ are ideals of R , and $\hat{R} = e\hat{R} \oplus (1 - e)\hat{R}$, $e\hat{R}$ is von Neumann along with $\hat{R}$, and it follows from 3, that $Z_r(eR) = 0$. Since $\hat{R}_R$ is injective, and since $(e\hat{R})_R$ is a direct summand of $\hat{R}_R$, it follows that $e\hat{R}$ is right self-injective, and then footnote (*) shows that $e\hat{R}$ is a maximal quotient ring of eR .

Assume now that R is semiprime and that N is a nilpotent right ideal of eR . Then $R \cap N$ is a nilpotent right ideal of R , and therefore $R \cap N = 0$. Since N is a right R-submodule of $\hat{R}$, it follows that $N = 0$, and eR is therefore semiprime.

9. SEMIPRIME RINGS WITH MAXIMUM CONDITION

1. PROPOSITION. Let S be a von Neumann ring with identity. Then the following properties are equivalent:

(1) The set $P_r(S)$ of principal right ideals of S satisfies the maximum condition; (2) $P_r(S)$ satisfies the minimum condition; (3) S is semisimple, artinian, and noetherian (both right and left).

PROOF. From §6 we know that if R is a ring, then R_R is semisimple if and only if each right ideal of R is a direct summand. If R has an identity, it is easy to see that a right ideal I is a direct summand of R if and only if there exists $e = e^2 \varepsilon R$ such that $I = eR$.

(1) → (2). Let J be any right ideal of S. By (1), J contains a principal right ideal I which is maximal in the set of those right ideals of $P_r(S)$ contained in J. By 4.14, the sum of two right ideals of $P_r(S)$ belongs to $P_r(S)$. This shows that $J = I$, so $J \varepsilon P_r(S)$. Since each principal right (resp. left) ideal of S is generated by an idempotent, by the remark made above, we conclude that S_S is semisimple, artinian, and noetherian. Hence $P_r(S)$ satisfies the minimum condition.

(2) → (3). The correspondence $Q \to Q^r$ (= the right annihilator of Q in S) is 1-1 and order-inverting between the set $P_l(S)$ of principal left ideals of S and the set $P_r(S)$. (In fact: if $Q = Se$, where $e = e^2 \varepsilon S$, then $Q^r = (1 - e)S$, and then $Q^{rl} = Q$.) Hence $P_l(S)$ satisfies the maximum condition. By the left-right symmetry of the proposition (1) → (2), we conclude that ${}_SS$ is semisimple, artinian and noetherian. Then $P_l(S)$ satisfies the minimum condition, so by symmetry again we deduce that S_S is semisimple, artinian and noetherian. Thus, (2) → (3). Clearly (3) → (1), and the proof is complete.

2. COROLLARY. Let R satisfy $Z_r(R) = 0$. Then the following statements are equivalent: (a) $C_r(R)$ satisfies the maximum condition; (b) $C_r(R)$ satisfies the minimum condition; (c) $\hat{R}$ is semisimple, artinian, and noetherian.

PROOF. Since $C_r(\hat{R}) \cong C_r(R)$, it suffices to assume that $R = \hat{R}$.

By 8.4, $C_r(R) = P_r(R)$. Since $\hat{R}$ is von Neumann, the corollary is now an immediate consequence of the proposition.

3. PROPOSITION. Let M_R be a module, and let $\lambda \in \Lambda = \text{Hom}_R(M,M)$. (1) If M_R is artinian, and if λ has a left inverse in Λ, that is, if λ is a monomorphism, then λ has a two-sided inverse in Λ, that is, λ is an automorphism. (2) If M_R is noetherian, and if λ is an epimorphism, that is, if $\lambda M = M$, then λ is an automorphism.

PROOF. (1) Since $M \supseteq \lambda M \supseteq \dots \supseteq \lambda^k M \supseteq \dots$ is a descending chain of submodules of M, there exists k such that $\lambda^k M = \lambda^{k+1} M$. If $\theta \in \Lambda$ is such that $\theta\lambda = 1$, then $\theta^k \lambda^k = 1$, and so $\theta^k \lambda^k M = M = \theta^k \lambda^{k+1} M = \lambda M$. Thus, λ is an automorphism.

(2) Since $\ker(\lambda) \subseteq \ker(\lambda^2) \subseteq \dots \subseteq \ker(\lambda^k) \subseteq \dots$ is an ascending sequence of submodules of M, there exists k such that $\ker(\lambda^k) = \ker(\lambda^{k+1}) = \dots = \ker(\lambda^{2k})$. Let $y \in \ker(\lambda^k)$. Since $\lambda^k M = M$, there exists $x \in M$ such that $\lambda^k x = y$. Then $\lambda^{2k} x = \lambda^k y = 0$, so $x \in \ker(\lambda^{2k}) = \ker(\lambda^k)$. Thus $y = \lambda^k x = 0$, and $\ker(\lambda^k) = 0$. Thus, $\ker(\lambda) = 0$, so λ is an automorphism.

4. COROLLARY. If R is a right artinian ring with identity, and if $x \in R$ is not a left zero divisor of R, then x has a (two-sided) inverse in R.

PROOF. R_R is artinian, and R is isomorphic to $\Lambda = \text{Hom}_R(R,R)$ under the correspondence $a \to a_L$, where $a_L \in \Lambda$ is the left multiplication of R performed by a. Since $xy = 0 \to y = 0 \ \forall \ y \in R$, x_L is a monomorphism of R_R, which has an inverse in Λ by the proposition. Then x has an inverse in R.

Let $A_r(R)$ denote the totality of annihilator right ideals of R. When $Z_r(R) = 0$, then $A_r(R) \subseteq C_r(R)$ by 8.5. When $C_r(R)$ is artinian, in particular where $\hat{R}$ is a right artinian ring, we see that R satisfies the minimum condition on annihilator right ideals. (Since $\hat{R}$ is then right noetherian, R also satisfies the maximum condition on annihilator (also closed) right ideals.)

A regular element in a ring R is an element which is neither a left nor a right zero divisor in R.

5. THEOREM. Let $Z_r(R) = 0$. Then: (1) R is semiprime and (2) $\hat{R}$ is (right) artinian if and only if (3): each essential right ideal I of R contains a regular element which has a two-sided inverse in $\hat{R}$.

PROOF. Since $S = \hat{R}$ is von Neumann, if S is right artinian, then S is semisimple, and therefore by §3,

$$S = S_1 \oplus \dots \oplus S_t ,$$

where S_i is a simple ring with minimum condition, $i = 1,\dots,t$. Let e_i denote the identity element of S_i, and set $R_i = e_iR$, $i = 1,\dots,t$. Since $e_1,\dots,e_t$ are central idempotents of S, and since $S_i = e_iS$, it follows from 8.11 that R_i is semiprime with quotient ring S_i, $i = 1,\dots,t$. Then, since S_i is prime, 8.10 implies that R_i is prime $\forall$ i.

Now let x^r denote the right annihilator in R of $x \in R$. Since $A_r(R)$ satisfies the minimum condition, if I is an essential right ideal of R, for each i, we can choose $x_i \in I \cap S_i$ such that x_i^r is minimal in $\{y^r \mid y \in I \cap S_i\}$.

We claim that x_i possesses a two-sided inverse in S_i. For if not, since S_i is simple and artinian, we can conclude from 3, that $x_iS_i \neq S_i$. Since S_i is semisimple, x_iS is a direct summand of S_i, so there exists a nonzero right ideal J_i of S_i such that $x_iS \cap J_i = 0$. (Note: J_i is also a right ideal of S, and $I \cap J_i$ is a right ideal of R_i.)

If $a \in I \cap J_i$, then (1) $(a + x_i)^r = a^r \cap x_i^r$. Clearly $(a + x_i)^r \supseteq a^r \cap x_i^r$. However, if $y \in (a + x_i)^r$, then $ay = -x_iy \in J_i \cap x_iS = 0$, and so $ay = x_iy = 0$, and $y \in a^r \cap x_i^r$. This establishes (1).

Since $a + x_i \in I \cap S_i$, and since x_i^r is minimal in $\{y^r \mid y \in I \cap S_i\}$, we see that $x_i^r = a^r \cap x_i^r = (a + x_i)^r \subseteq x_i^r$, so $a^r \cap x_i^r = x_i^r$ and $x_i^r \subseteq a^r$. Thus $ax_i^r = 0 \ \forall \ a \in I \cap J_i$. Then we have $(I \cap J_i)x_i^r = 0$, and so $(I \cap J_i)(e_ix_i^r) = 0$ (since $(I \cap J_i)e_i = I \cap J_i$). Since I is an essential R-submodule of S, necessarily $I \cap J_i \neq 0$. Then, since $I \cap J_i \subseteq R_i$, $e_ix_i^r = 0$ by primeness of R_i. Thus, x_i is not a left divisor of zero in R_i, and it follows that x_i is not a left divisor of zero in the quotient ring S_i of R_i. Thus, by 4, we conclude that x_i has a two-sided inverse y_i in S_i , $i = 1,\dots,t$. Then $y = y_1 + \dots + y_t$ is a two-sided inverse of $x = x_1 + \dots + x_t \in I$. This proves the necessity.

Conversely, assume (3) and N be any nilpotent ideal of R . If $N \neq 0$, let $U = N^{k-1}$, where k = index of nilpotency of N . Let K be a complement right ideal of R corresponding to U . Then $U + K$ is an essential right ideal of R , so $U + K$ contains an element x such that $x^{-1} \in \hat{R}$. But $KU \subseteq K \cap U = 0$, so $(U + K)U = 0$. Then $xU = 0$ and $U = 0$. Thus $N^{k-1} = 0$, a contradiction. Hence $N = 0$, and (1) holds.

Now let I be any right ideal of $\hat{R}$, and let K be a complement right ideal of $\hat{R}$ corresponding to I . Then $J = I + K$ is an essential right ideal of $\hat{R}$. Assume for the moment that $Q = J \cap R$ is an essential right ideal of R . Then there would exist $x \in Q$ such that $x^{-1} \in \hat{R}$. Then $1 = xx^{-1} \in J\hat{R} \subseteq J$, so $J = \hat{R} = I \oplus K$. Then each right ideal of $\hat{R}$ is a direct summand, and $\hat{R}_{\hat{R}}$ is accordingly semisimple. Since $1 \in \hat{R}$, $\hat{R}$ is (right) artinian.

Hence we need only verify the implication: if J is an essential right ideal of R , then $Q = J \cap R$ is an essential right ideal of R . Clearly Q is a right ideal of R . Suppose P is a right ideal of R such that $Q \cap P = 0$. Then $J \cap P = 0$, and hence $J \cap P^* = 0$, where P^* is the injective hull of P_R taken in $\hat{R}_R$. Since P^* is a right ideal of $\hat{R}$ (see §8), and since J is an essential right ideal of $\hat{R}$, we conclude that $P^* = 0$, whence $P = 0$, and Q is therefore an essential right ideal of R .

A ring S is a classical right quotient ring of a subring R in case: (1) S contains an identity element; (2) each regular element of R has a two-sided inverse in S ; (3) $S = \{ab^{-1} \mid a, \text{ regular } b \in R\}$.

6. LEMMA. Let Q be a classical quotient ring of a ring R . Then: (1) if $b_1,\dots,b_n$ are regular elements of R , there exist a regular element $c \in R$ and elements $g_i \in R$ such that $b_i^{-1} = g_i c^{-1}$, $i = 1,\dots,n$; (2) if $x_1,\dots,x_n \in Q$, there exists a regular element $c \in R$ such that $x_i c \in R$, $i = 1,\dots,n$; (3) if I is a right ideal of R , then the right ideal of Q generated by I is IQ , and $IQ = xc^{-1} \mid x \in I$, regular $c \in R\}$. (4) If $d \in R$ is regular, then dR is an essential right ideal of R.

PROOF. (1) is true for $n = 1$, since then $g_1 = b_1$, $c_1 = b_1^2$ will work. Assume the theorem for $b_1, \dots, b_{n-1}$, and let $\bar{g}_i \in R$, regular $\bar{d} \in R$ be such that $b_i^{-1} = \bar{g}_i \bar{d}^{-1}$, $i = 1,\dots,n-1$. Now there exist g_n, regular $d \in R$ such that $b_n^{-1}\bar{d} = g_n d^{-1}$. Hence $c = \bar{d}d$, $g_i = \bar{g}_i d$, $i = 1,\dots,n-1$, are the desired elements.

(2) Write $x_i = a_i b_i^{-1}$, with $a_i, b_i \in R$, and choose c, g_i in accordance with (1), $i = 1,\dots,n$. Then $x_i c = a_i b_i^{-1} c = a_i g_i \in R$, $i = 1,\dots,n$, and c is regular in R.

(3) Clearly IQ is the right ideal of Q generated by I, and $IQ \supseteq I_0 = \{xc^{-1} \mid x \in I$, regular $c \in R\}$. A typical element $y \in IQ$ has the form $y = \Sigma_{i=1}^n y_i x_i$, with $y_i \in I$, $x_i \in Q$, $i = 1,\dots,n$, and n an integer. By (2), there exists a regular $c \in R$ such that $x_i c = g_i \in R$. Then $y = xc^{-1}$, where $x = \Sigma_1^n y_i g_i \in I$. Thus $IQ = I_0$.

(4) Let $I = dR$. Since $1 = dd^{-1} \in IQ$, necessarily $Q = IQ$. If K is any nonzero right ideal of R, and if $0 \neq k \in K$, then (3) implies that $k = xc^{-1}$, with $x \in I$, $c \in R$. Then clearly $kc = x \neq 0$, and so $I \cap K \neq 0$. This proves (4).

We say R is right quotient-semisimple (resp. quotient-simple) in case R possesses a classical right quotient ring S which is semisimple (resp. simple) artinian.

7. THEOREM. A ring R is right quotient-semisimple (resp. quotient-simple) if and only if the following conditions are satisfied:

(1) $Z_r(R) = 0$

(2) $C_r(R)$ satisfies the maximum condition

(3) R is semiprime (resp. prime).

PROOF. Let R have classical right quotient ring S which is semisimple. Clearly S_R is an essential extension of R_R. Since S is a von Neumann ring, by 8.3, we deduce that $Z_r(R) = 0$, proving (1). Since S_S is injective, it follows that $S = \hat{R}$. Then (2) holds by 2.

(3). Let U be a nilpotent ideal of R. If $U \neq 0$, $N = U^{k-1}$ satisfies $N^2 = 0$, where k = index of nilpotency of U. Now SNS is an ideal of S. Since S_S is semisimple, it follows that $SNS = eS$, where e is an idempotent lying in the center of S. By 6, $NS = \{xc^{-1} \mid x \in N$, regular $c \in R\}$. Hence, since $e \in SNS$, there exist elements

$s_i \in S$, $x_i \in N$, regular $c_i \in R$ such that $e = \Sigma_1^n s_i x_i c_i^{-1}$. Again by 6, there exists regular $c \in R$ such that $c_i^{-1}c = g_i \in R$, $i = 1,\ldots,n$. Then $ec = \Sigma_1^n s_i x_i g_i \in SN$. It follows that $ecN^{k-1} = 0$. Since $ec = ce$, we obtain $eN^{k-1} = 0$. Since $ey = y \;\forall\; y \in eS$, and since $N \subseteq SNS = eS$, we conclude that $N^{k-1} = eN^{k-1} = 0$, a contradiction proving (3).

Now assume (1) - (3). (2), together with 2, implies that $\hat{R}$ is semisimple artinian. This fact, together with (3) and 5, implies that each essential right ideal of R contains an element which has an inverse in $\hat{R}$. But if $a \in \hat{R}$, then $(R:a)$ is an essential right ideal. Hence there exists $x \in R$ such that $x^{-1} \in R$ and such that $ax = r \in R$. Then $a = rx^{-1}$. In order to show that $\hat{R}$ is a classical right quotient ring of R, it remains only to show that each regular element $c \in R$ has an inverse in $\hat{R}$. To do this, note that c is not a left divisor of zero in $\hat{R}$. For suppose $0 \neq y \in \hat{R}$ is such that $cy = 0$. Then $y = ax^{-1}$ with $a, x \in R$, so $cax^{-1} = ca = 0$, and $0 \neq a \in R$, a contradiction to the fact that c was assumed to be regular in R. Now we apply the fact that each non-left zero divisor of a semisimple artinian ring $\hat{R}$ has an inverse in $\hat{R}$. Hence $c^{-1} \in \hat{R}$, completing the proof of the semisimple case.

If $\hat{R}$ is simple, then $\hat{R}$ is prime, and then R is prime by 8.10. If R is prime, then $\hat{R}$ is prime by 8.10, and then $\hat{R}$ is simple. This completes the proof.

10. NIL AND SINGULAR IDEALS UNDER MAXIMUM CONDITIONS

In this section we prove a theorem of Levitzki on the nilpotency of nil ideals in a noetherian ring. The proof we give is a short one by Y. Utumi [Amer. Math. Monthly, vol.70 (1963) p.286]. We then apply this result to derive Goldie's characterization of quotient-semisimple rings.

This first two lemmas hold also in the case S is a semigroup. Recall that a (right, left) ideal I of S is nil in case each element of I is nilpotent.

1. LEMMA. Let R be any ring. Then R contains no nil left ideals $\neq 0$ if and only if R contains no nil right ideals $\neq 0$.

PROOF. Let I be a nonzero nil right ideal, and let $0 \neq a \in I$. If $Ra = 0$, then $a \in R^r = \{x \in R \mid Rx = 0\}$, so R^r is a nonzero nilpotent ideal of R. If $Ra \neq 0$, and if $u \in R$, then $au \in I$ au is nilpotent. Since $(ua)^{n+1} = u(au)^n a$, ua is also nilpotent, i.e., Ra is a nonzero nil left ideal.

The following question is open: If S contains a nil left ideal $\neq 0$, does S contain a nil ideal $\neq 0$?

2. LEMMA. Let S be a ring satisfying the maximum condition for annihilator right ideals. If S has a nonzero nil right or left ideal A, then S contains a nonzero nilpotent ideal.

PROOF. By 1, we can assume A is a nil left ideal $\neq 0$. Let $0 \neq a \in A$ be such that a^r is maximal in $\{x^r \mid 0 \neq x \in A\}$. If $u \in S$ is such that $ua \neq 0$, then $(ua)^n = 0$ and $(ua)^{n-1} \neq 0$ for some $n > 1$. Since $[(ua)^{n-1}]^r \supseteq a^r$, and since $ua \in A$, it follows that $[(ua)^{n-1}]^r = a^r$. Since $ua \in [(ua)^{n-1}]^r$, then $aua = 0$. Thus $aSa = 0$, and then $(a)^3 = 0$, where (a) is the ideal of S generated by a.

3. THEOREM OF LEVITZKI. If S is a right noetherian ring, then every nil left or right ideal I of S is nilpotent.

PROOF. Let N be the maximal nilpotent ideal of S. Then the difference ring $S - N$ contains no nilpotent ideals $\neq 0$. If I is not contained in N, $(I + N) - N$

would be a nonzero nil onesided ideal of $S - N$. But $S - N$ is right noetherian, so $(I + N) - N$ would be a nilpotent ideal by 2. This contradiction establishes that $I \subseteq N$, so I is nilpotent.

As before, $A_r(R)$ denotes the set of annihilator right ideals of R.

4. PROPOSITION. If $A_r(R)$ satisfies the maximum condition, then $Z_r(R)$ is nil ideal.

PROOF. If $x \in R$, then

$$x^r \subseteq (x^2)^r \subseteq \dots \subseteq (x^n)^r \subseteq \dots$$

is an ascending sequence of ideals of $A_r(R)$, so by hypothesis there exists n such that

$$(x^n)^r = (x^{n+1})^r = \dots = (x^{2n})^r = \dots$$

Hence $y = x^n$ satisfies $y^r = (y^2)^r$. Now the left annihilator ideal R^l satisfies $(R^l)^2 = 0$. Hence if x is non-nilpotent, then $y \notin R^l$, so $yR \neq 0$. Suppose for the moment that $x \in Z_r(R)$ is not nilpotent. Then $y = x^n \in Z_r(R)$ and so $y^r \cap yR \neq 0$. Let $t \in R$ be such that $0 \neq yt \in y^r \cap yR$. Then $y(yt) = y^2t = 0$, so $t \in (y^2)^r$. But $(y^2)^r = y^r$, so $yt = 0$, a contradiction showing that each $x \in Z_r(R)$ is nilpotent. Thus $Z_r(R)$ is a nil ideal.

2. and 4. have the following consequence:

5. COROLLARY. If R is semiprime, and if $A_r(R)$ satisfies the maximum condition, then (1) R contains no nil one-sided ideals $\neq 0$, and (2) $Z_r(R) = 0$.

Let (1r) denote the condition that the complement (= closed) right ideals of R satisfy the maximum condition, and let (2r) denote the condition that the annihilator right ideals of R satisfy the maximum condition.

6. THEOREM. (Goldie [2]). A ring R is right quotient-semisimple if and only if R is a semiprime ring satisfying (1r) and (2r).

PROOF. If R is right quotient-semisimple, then (1) - (3) of Theorem 9.7 hold. Hence R is semiprime, $Z_r(R) = 0$, and (1r) holds. By 8.5, $Z_r(R) = 0$ implies that each annihilator right ideal is a complement right ideal, so we conclude that (2r) holds.

Conversely, let R be a semiprime ring satisfying (1r) and (2r). Then 5 implies that $Z_r(R) = 0$, and (1r) implies that $C_r(R)$ satisfies the maximum condition. By 9.7, R is right quotient-semisimple.

Lesieur-Croisot [1] independently proved the following special case (cf. 9.7).

7. COROLLARY. R is right quotient-simple if and only if R is a prime ring satisfying (1r) and (2r).

8. COROLLARY. If R is a right noetherian prime (resp. semiprime) ring, then R is right quotient-simple (resp. semisimple).

An integral domain $R \neq 0$ is a <u>right Ore domain</u> in case R has a right quotient field. The corollary shows that each right noetherian integral domain is a right Ore domain.

DEFINITION. Let R be right quotient-semisimple, and let Q be its right quotient ring. Write $Q = Q_1 \oplus \dots \oplus Q_t$, where Q_i are the unique simple ideals of Q, $i = 1,\dots,t$. Since Q_i is a simple artinian ring, Q_i is isomorphic to a full ring of $n_i \times n_i$ matrices over a field; let $d(Q_i) = n_i$, $i = 1,\dots,n$. Then $d(Q_i)$ is uniquely determined by Q_i, and $\Sigma_1^t d(Q_i)$ is uniquely determined by Q. We let d(Q) denote this integer. Since Q is uniquely determined (up to isomorphism) by R, we can unambiguously define dim R = d(Q), and denote this integer by d(R).

If $Q = D_n$, where D is a field, and if I_i denotes the set of all matrices with zeros off the i-th row and arbitrary entries from D in the i-th row, then it is easily checked that I_i is a minimal right ideal of Q, and that $Q = I_1 \oplus \dots \oplus I_n$. Thus, Q is a direct sum of n = d(Q) right minimal right ideals.

Next if Q is semisimple artinian, as in the definition, then it follows that (1) Q is a direct sum of d(Q) minimal right ideals, and by the Jordan-Holder theorem, (2) each direct sum of minimal right ideals has length $\leq d(Q)$. These facts will be used in the proof of the next result.

9. **PROPOSITION.** Let R be right quotient-semisimple, and let $n = \dim R$. Then each family of independent right ideals of R has cardinality $\leq n$, and R contains a family of n independent right ideals.

PROOF. Let Q be the quotient ring of R, and let $\{K_i \mid i \in I\}$ be an independent family of right ideals of R. By the proof of 8.4, the injective hull K_i^* of $(K_i)_R$ taken in Q_R is a right ideal of Q $\forall$ $i \in I$, and by 3.6, $\{K_i^* \mid i \in I\}$ is an independent family of right ideals of Q. Since Q is right artinian, each K_i^* contains a minimal right ideal J_i of Q, and the family $\{J_i \mid i \in I\}$ is independent. By (ii) above, card $I \leq n = \dim R = d(Q)$. By (i), Q contains n independent minimal right ideals $T_1, \ldots, T_n$, and then $T_1 \cap R, \ldots, T_n \cap R$ are independent right ideals of R.

11. STRUCTURE OF NOETHERIAN PRIME RINGS

The Wedderburn-Artin Theorem (3.21) implies that a right artinian prime ring S is a simple finite dimensional full ring: $S = D_n$, where D is a field. Such a ring is right noetherian. The first general result on the structure of non-commutative noetherian prime (or simple) rings was Goldie's Theorem (Goldie [1]) which states that if R is a prime ring which is both right and left noetherian, then R is both left and right quotient-simple with classical quotient ring $S = D_n$, D a field. Later Goldie [2], and Lesieur-Croisot [1], generalized this result to right noetherian prime rings: each such ring R is right quotient-simple (10.8). This established for the first time a connection between general right noetherian prime rings and right artinian prime rings.

As close as this connection is, it yields very little information on the internal structure of R (cf. 9.7). For example, it is not immediately obvious that R contains nonzero nilpotent elements when R is not an integral domain (see below).

By imposing a stronger chain condition, Goldie [3] obtained a decisive result: If R is a prime ring with identity which is a principal right ideal ring, then $R \cong F_n$, where F is a right Ore domain.

The main result of this section states that a right noetherian prime ring R always contains a subring $\cong F_n$, where F is a right Ore domain with quotient field D such that D_n is the right quotient ring of R . (See Theorem 6 for more precise result.) Although we can easily deduce from this that R contains nonzero nilpotent elements when R is not an integral domain, R itself may fail to contain non-trivial idempotents even when R contains an identity. (See Supplementary Remark C). Nevertheless, Goldie's Theorem on principal right ideal prime rings is an easy consequence of the proof of our theorem. (See Corollary 7)

If A is a ring containing R , then as before A is a <u>right quotient ring of</u> R in case $(A \nabla R)_R$. Any classical right quotient ring S of a ring R is a right quotient ring in this sense, so the first three results of this section are applicable to classical quotient rings.

1. **THEOREM.** Let R be a semiprime ring, and let A be a right quotient ring of R. Let e be an idempotent of A such that $D = eAe$ is a field. If $K = eAe \cap R \neq 0$, then K is a right Ore domain, and D is its right quotient field, $D = \{kq^{-1} \mid k, 0 \neq q \in K\}$.

PROOF. First note that eA is a left vector space over $D = eAe$. Thus if $0 \neq d \in D$, then $dx = 0 \Rightarrow x = 0 \ \forall \ x \in eA$. Let $[d]$ denote the R-submodule of A generated by $d \in D$. If we set $d(r,n) = dr + nd \ \forall \ (r,n) \in R \times Z$ (cartesian product) $r \in R$, $n \in Z$, then $[d] = \{d(r,n) \mid (r,n) \in R \times Z\}$. Since $(A \nabla R)_R$, we have $[d] \cap R \neq 0 \ \forall \ 0 \neq d \in D$. Hence let $0 \neq c = d(r,n) \in [d] \cap R$. Since $d = de$, clearly $e(r,n) \neq 0$, so choose $(r',n') \in R \times Z$ such that $0 \neq e(r,n)(r',n') \in R$. Then setting $U = eA \cap R$, and setting $a = e(r,n)(r',n')$, we see that $a \in U$ and $b = da = c(r',n') \in U$. Since $a \neq 0$, necessarily $b \neq 0$. Since $V = Ae \cap R$ is a left ideal of R, $K = eAe \cap R = V \cap U$ is a left ideal of U. Furthermore, since $K \neq 0$ by hypothesis, since $K \subseteq D$, and since $U \subseteq eA$, U is a torsion-free left K-module. Since R is semiprime, the left annihilator ideal of R is zero, so $aR \neq 0$. Since $aR \subseteq U$, it follows that $KaR \neq 0$. Since KaR is a right ideal of R, semiprimeness of R implies that $(KaR)^2 \neq 0$, hence $aRK \neq 0$. Now choose $t \in R$, $q \in K$ such that $x = atq \neq 0$. Since $x \in eA$, necessarily $y = dx \neq 0$. Since $at, bt \in U$ (U is a right ideal of R and $a,b \in U$), necessarily $x = atq$, $y = btq \ K \in (K$ is a left ideal of U and $t,q \in K)$. If x^{-1} denotes the inverse of x in D, we see that $d = yx^{-1}$, and $x,y \in K$. Thus, D is the right quotient field of K, completing the proof.

REMARK. Any quotient ring of a semiprime ring is semiprime. Thus (see Jacobson [1, p.65, Prop. 1]) eAe is a field if and only if eA is a minimal right ideal.

Below we show that the vanishing of $Z_r(Q)$ is enough to insure transitivity of the relation "quotient ring of". In the proof, if x is an element in the ring Q, and if P is a subring of Q, then

$$(P:x) = \{p \in P \mid xp \in P\}$$

is a right ideal of P. Furthermore, if $(Q \nabla P)_P$, then $(P:x)$ is an essential right ideal of P. (See §5).

2. LEMMA. Let Q be a right quotient ring of R, and let R be a right quotient ring of T. (1) If I is any right ideal of Q such that $I \cap R$ is an essential right ideal of R, then I is an essential right ideal of Q; (2) $Z_r(Q) \supseteq Z_r(R) \supseteq$ $\supseteq Z_r(T)$; (3) If $Z_r(Q) = 0$, then Q is a right quotient ring of T.

PROOF. (1) is trivial. (2) Let x^r denote the right annihilator in Q of $x \in Q$. If $x \in Z_r(R)$, then $x^r \cap R$ is an essential right ideal of R . Then (1) implies that $x \in Z_r(Q)$, proving (2).

(3) If $x \in Q$, $xR = 0$ implies by (1) that $x \in Z_r(Q)$. Since $Z_r(Q) = 0$, if $0 \neq x \in Q$, then $xR \neq 0$, so $xR \cap R \neq 0$. Let $x,r \in R$ be such that $s = xr \neq 0$. Now $(T:r)$ (resp. $(T:s)$) is an essential right ideal of T , and so is $(T:r) \cap (T:s)$. Hence $s^r \supseteq (T:r) \cap (T:s)$ would imply by (1) that $s \in Z_r(R)$. But $Z_r(R) = 0$ by (2) and $s \neq 0$, so we conclude that $s^r \not\supseteq (T:r) \cap (T:s)$. Accordingly we can choose $t \in (T:r) \cap (T:s)$ such that $st \neq 0$. Then $st = x(rt) \in xT \cap T$, so $xT \cap T \neq 0$. This proves (3).

For convenience, we recall the definition of a prime ring. R is said to be prime in case any of the following three equivalent conditions are satisfied:

(a) $I^r = 0 \quad \forall$ right ideals I ;

(b) $I^l = 0 \quad \forall$ left ideals I ;

(c) $xRy = 0 \rightarrow x = 0$ or $y = 0, \quad \forall\, x,y \in R$.

Here $I^r = \{a \in R \mid Ia = 0\}$, and $I^l = \{a \in R \mid aI = 0\}$.

If A (resp. B) is a left (resp. right) ideal of R, then $T = BA$ is defined to be the set of all finite sums of the products ba, $b \in B$, $a \in A$. It is to be observed that T is a subring of R .

3. PROPOSITION. Let $R \neq 0$ be a prime ring, let A be a left ideal of R whose right annihilator A^r in R is zero, and let B be a right ideal of R whose left annihilator B^l in R is zero. Then: (1) $T = BA$ is a prime ring; (2) If, in addition, B is an essential right ideal of R , then R is a right quotient ring of T .

PROOF. (1) Let $x,y \in T$ be such that $xTy = 0$. Then $AyRxB$ is an ideal of R having square zero, so primeness of R yields $AyRxB = 0$. Then again by primeness of R, $Ay = 0$, or $xB = 0$. Since $A^r = B^l = 0$, we obtain $y = 0$ or $x = 0$, and T is therefore prime.

(2) If $0 \neq x \in R$, then $xB \neq 0$. Then $(R \nabla B)_R$ implies $xB \cap B \neq 0$. Let $b \in B$ be such that $0 \neq xb \in B$. Primeness of R implies $A^l = 0$, so $xbA \neq 0$. But $bA \subseteq T$, and $xbA \subseteq T$, so $xT \cap T \neq 0$, proving (2).

4. PROPOSITION. Let Q be a classical right quotient ring of R. If R has a classical left quotient ring S, then Q is a classical left quotient ring of R.

PROOF. Let $a \circ b$ denote multiplication in S, and let ab indicate that of Q. If a, regular $b \in R$, then $a \circ b^{-1} \in S$, so there exist c, regular $d \in R$ such that $a \circ b^{-1} = d^{-1} \circ c$, that is, such that $da = cb$. Then $ab^{-1} = d^{-1}c$ in Q.

Let $P = \{x^{-1}y \in Q \mid y$, regular $x \in R\}$. If $x^{-1}y$, $u^{-1}v \in P$, then we have just shown that there exist $c,d \in R$ such that $yu^{-1} = d^{-1}c$. Then $(x^{-1}y)(u^{-1}v) = (dx)^{-1}(cv) \in P$, and P is therefore closed under multiplication. Similarly, P is closed under addition, so P is a subring of Q. Clearly $P = Q$, so Q is a classical left quotient ring.

A consequence of this proposition is that if a ring R is right quotient-simple (resp. semisimple) with quotient ring Q, then R is left quotient-simple (resp. semisimple) if and only if Q is a classical left quotient ring of R. We use this fact without explicit mention in the proof of (2) of our main result (Theorem 6).

If A is a ring with identity 1, a subset $\{e_{ij} \mid i,j = 1,\dots,n\}$ of A is a set of matrix units in case

$$(1) \quad e_{ij}e_{pq} = \delta_{jp}e_{iq} \qquad i,j,p,q = 1,\dots,n,$$

where δ_{jp} is the Kronecker-δ, and (2) $\Sigma_{k=1}^{n} e_{kk} = 1$.

If $A = B_n$ is the ring of $n \times n$ matrices over a ring B having identity, and if e_{ij} denotes the matrix of A having 1 in the (i,j)-position, and 0's elsewhere, one easily verifies that $\{e_{ij} \mid i,j = 1,\dots,n\}$ is a set of matrix units. The next result shows, conversely, that the existence of a set of matrix units in a ring implies that the ring is a matrix ring.

5. PROPOSITION. Let $\{e_{ij} \mid i,j = 1,\ldots,n\}$ be a set of matrix units in a ring A with identity 1, and let B be the subring of A consisting of all elements which commute with the $e_{ij}, i,j = 1,\ldots,n$. Then every element of A can be written in one and only one way as $\Sigma_{i,j=1}^{n} b_{ij}e_{ij}$, where $b_{ij} \in B$, $i,j = 1,\ldots,n$. Hence $A \cong B_n$. The ring B is isomorphic to $e_{11}Ae_{11}$.

PROOF. If $a \in A$, define $a_{ij} = \Sigma_{k=1}^{n} e_{ki}ae_{jk}$, $i,j = 1,\ldots,n$. Then

$$a_{ij}e_{pq} = e_{pi}ae_{jq} = e_{pq}a_{ij}, \quad i,j = 1,\ldots,n.$$

Thus, $a_{ij} \in B \ \forall \ a \in A, i,j = 1,\ldots,n$.

In particular, $a_{ij}e_{ij} = e_{ii}ae_{jj}$, $i,j = 1,\ldots,n$. Hence

$$\Sigma_{i,j=1}^{n} a_{ij}e_{ij} = \Sigma_{i,j=1}^{n} e_{ii}ae_{jj} = a$$

If b_{ij} are elements of B such that $\Sigma_{i,j=1}^{n} b_{ij}e_{ij} = 0$, then

$$0 = \Sigma_{k=1}^{n} e_{kp}\left(\Sigma_{i,j=1}^{n} b_{ij}e_{ij}\right)e_{qk} = b_{pq} = 0, \quad p,q = 1,\ldots,n.$$

This proves the first assertion. This implies the mapping $\Sigma_{i,j=1}^{n} b_{ij}e_{ij} \to$ the matrix with b_{ij} in the (i,j)-position is 1-1 onto B_n. One verifies that this is a ring isomorphism, $A \cong B_n$. Now $e_{11}Ae_{11} = \{be_{11} \mid b \in B\}$, and clearly $b \to be_{11}$ is a ring isomorphism of B onto Be_{11}. Hence $B \cong e_{11}Ae_{11}$.

If $Q = D_n$, where D is a field, then by 5, there exists a set $M = \{e_{ij} \mid i,j = 1,\ldots,n\}$ of matrix units of Q, and the set of elements of Q which commute with M is a field isomorphic to D. Without loss of generality we can assume this field is D, and call it the centralizer of M in Q. By 5, $Q = \Sigma_{i,j=1}^{n} De_{ij}$. If x is any invertible element of Q, then $N = x^{-1}Mx$ is a set of matrix units of Q whose centralizer is $x^{-1}Dx$. We call any such set of matrix units of Q a <u>complete</u> or <u>full set of matrix units</u>.

6. THEOREM. Let R be a right quotient-simple ring with quotient ring $Q = D_n$, D a field. (1) Then Q contains a complete set $M = \{e_{ij} \mid i,j = 1,\dots,n\}$ of matrix units with the following property: if D is the centralizer of M in Q, then R contains a subring

$$F_n = \Sigma_{i,j=1}^{n} Fe_{ij} ,$$

where F is a right Ore domain contained in $R \cap D$ and D is the right quotient field of F. Furthermore:

$$Q = \{ak^{-1} \mid a \in F_n , 0 \neq k \in F\} .$$

(2) If R is also left quotient-simple, then every complete set M of matrix units has the property described in (1), and each corresponding D is also the left quotient field of F. Finally,

$$Q = \{q^{-1}b \mid b \in F_n , 0 \neq q \in F\} .$$

PROOF. We give a proof of (1) and (2) simultaneously by showing if $M = \{e_{ij} \mid i,j = 1,\dots,n\}$ is any complete set of matrix units of Q such that

(*) there exists a regular element $y \in R$ such that $yM \subseteq R$

then M has the property in statement (1).

Now if R is also left quotient-simple, then Q is a classical left quotient ring of R, and the right-left symmetry of (2) of 9.6 asserts that each full set M has property (*).

Next assume only that R is right quotient-simple, and let N be a complete set of matrix units of Q. Then, by 9.6, there exists a regular $y \in R$ such that $Ny \subseteq R$. Hence, $M = y^{-1}Ny$ is a complete set of matrix units of Q satisfying (*).

Accordingly let $M = \{e_{ij} \mid i,j = 1,\dots,n\}$ be any complete set of matrix units of Q satisfying (*). Then, by 9.6, there exists regular $x \in R$ such that $Mx \subseteq R$. Hence, the left ideal $A = \{r \in R \mid rM \subseteq R\}$ contains the regular element $y \in R$, and the right ideal $B = \{r \in R \mid Mr \subseteq R\}$ contains the regular element $x \in R$. Furthermore,

B is an essential right ideal of R by (4) of 9.6.

Since R is a prime ring by 9.7, we apply 3 to conclude that $T = BA$ is a prime ring and that R is a right quotient ring of T. Since $Z_r(Q) = 0$, we deduce from (3) of 2 that Q is a right quotient ring of T.

Next we show that $e_{11}Qe_{11} \cap T \neq 0$. Now $0 \neq ye_{11} \in R$ and

$$ye_{11}e_{ij} = \begin{cases} ye_{ij}, & i = 1, \\ 0, & i \neq 1, \end{cases}$$

which shows that $ye_{11} \in A$. Since $x \in B$ and since x is regular, it follows that $0 \neq xye_{11} \in T = BA$, so that $T \cap Qe_{11} \neq 0$. Since Q is a right quotient ring of T, $e_{11}Q \cap T \neq 0$. Then primeness of T implies that $(e_{11}Q \cap T)c \neq 0$, where $c = xye_{11}$. If $d \in e_{11}Q \cap T$ is such that $dc \neq 0$, then $dc \in T \cap e_{11}Qe_{11}$, proving our assertion.

Since $F_1 = e_{11}Qe_{11} \cap T \neq 0$, and since $e_{11}Qe_{11}$ is a field ($\cong D$), 1 implies that $e_{11}Qe_{11}$ is the right quotient field of $F_1 = e_{11}Qe_{11} \cap T$. Since D is isomorphic to $De_{11} = e_{11}Qe_{11}$ under the map $\varphi: d \to de_{11}$, $d \in D$, this shows that D is the right quotient field of $F = \varphi^{-1}F_1$. Furthermore,

$$Fe_{ij} = e_{i1}F_1e_{1j} \subseteq e_{i1}Te_{ij} = (e_{i1}B)(Ae_{1j}) \subseteq RR \subseteq R,$$

$i,j = 1,\ldots,n$. Thus, R contains the subring

$$F_n = \Sigma_{i,j=1}^n Fe_{ij},$$

and $F \subseteq F \cap D$.

If $a = \Sigma_{i,j=1}^n e_{ij}d_{ij} \in Q$, $d_{ij} \in D$, $i,j = 1,\ldots,n$, then by 9.6, there exists $0 \neq k \in F$ such that $d_{ij}k = q_{ij} \in F$, $i,j = 1,\ldots,n$. Then $a = fk^{-1}$, where $f = \Sigma_{ij=1}^n e_{ij}q_{ij} \in F_n$. This proves (1).

If R is also left quotient-simple, then Q is the classical left quotient ring of R, and 1 implies that $e_{11}Qe_{11}$ (resp. D) is the left quotient field of F_1 (resp. F). The computation above establishes that if $a \in Q$, then $a = q^{-1}b$, with $b \in F_n$, and $0 \neq q \in F$. This completes the proof of (2).

8. COROLLARY. (A.W. Goldie [3]). If R is a principal right ideal ring, and if R is prime, then $R = K_n$, where K is a right Ore domain.

PROOF. Using the notation of the theorem, we can write $B = cR$ for some $c \in R$. If $b \in Q$ is such that $bc = 0$, then $bB = 0$. But $x \in B$ is regular in R , so $x^{-1} \in Q$. Thus $bx = 0$ and $b = 0$, so c is not a right zero divisor in Q. Then, as is well known in artinian rings, $c^{-1} \in Q$.

Trivially $e_{ij}B \in B$, that is, $e_{ij}cR \subseteq cR$ and $c^{-1}e_{ij}c = f_{ij} \in R$, $i,j = 1,\dots,n$. If G is the centralizer of $N = c^{-1}Mc$ in Q , it follows that

$$R = \Sigma_{i,j=1}^{n} Kf_{ij} ,$$

where $K = G \cap R$. Since Q is the classical right quotient ring of R , an easy computation shows that G is the right quotient field of K. (This fact also follows from the theorem since the theorem states that G is the right quotient field of some integral domain contained in K.)

At present (see Goldie [3]) it is unknown whether or not K has to be a principal right ideal domain.*) R. Bumby has shown me that the answer is "yes" if K is commutative.

9. COROLLARY. The class of right quotient-simple rings coincides with the class of intermediate rings of the extensions $(\hat{K}_n)$ over K_n , where K ranges over all right Ore domains, and n ranges over all natural numbers.

PROOF. Class inclusion one way is an immediate consequence of the theorem, and the reverse inclusion is a simple exercise which we omit.

SUPPLEMENTARY REMARKS. A. Let R be right quotient-simple with right quotient ring $Q = D_n$, $n > 1$. If f is any idempotent of Q such that fQ is a minimal right ideal, then $f = ac^{-1}$, with a, regular $c \in R$. Thus $e = c^{-1}fc$ is idempotent and $0 \neq ce \in Qe \cap R$. Since $Qe \cap R \neq 0$, it follows from primeness of R that $(eQ \cap R)(Qe \cap R) \neq 0$, so that $eQe \cap R \neq 0$. Then, Theorem 1.1 implies that $K = eQe \cap R$ is a right Ore domain, and $\hat{K} = eQe$ is its right quotient field. Since $\hat{K} \cong D$, we obtain that $Q = \hat{K}_n$, where K is a right Ore domain contained in R . This illustrates

*) (Added 1967) K need not be principal.

the precise nature of Theorem 6 which states much more.

B. Next we show that (2) of Theorem 6 fails without the hypothesis that R is also left quotient-simple. The example below was suggested by S.U. Chase.

Let K be a right Ore domain which is not a left Ore domain. Let x,y be nonzero elements of K such that $Kx \cap Ky = 0$, and let

$$R = \begin{pmatrix} Kx, & Ky \\ Kx, & Ky \end{pmatrix}$$

(R is the ring of all 2×2 matrices $\begin{pmatrix} ab \\ cd \end{pmatrix}$ with $a, c \in Kx$, and $b, d \in Ky$.) Since K is right Ore, if $A = \begin{pmatrix} ab \\ cd \end{pmatrix}$ is an arbitrary element of $\hat{K}_2$, there exists $0 \neq q \in K$ such that $aq, bq, cq, dq, \in K$, and then

$$B = \begin{pmatrix} ab \\ cd \end{pmatrix} \begin{pmatrix} qx & 0 \\ 0 & qy \end{pmatrix} \in R$$

Thus, $A = BC^{-1}$, with $B, C = \begin{pmatrix} qx & 0 \\ 0 & qy \end{pmatrix} \in R$. Hence, $\hat{K}_2$ is the classical right quotient ring of R, that is, R is right quotient-simple.

As in Theorem 6, we identify $\hat{K}$ with the subring of $\hat{K}_2$ consisting of all scalar matrices $\begin{pmatrix} k0 \\ 0k \end{pmatrix}$ with $k \in \hat{K}$. Now assume for the moment that R contains a subring F_2 where F is an integral domain $\subseteq \hat{K}$. The contradiction is immediately evident (even without assuming that $\hat{F} = \hat{K}$), since the form of R,

$$R = \begin{pmatrix} Kx, & Ky \\ Kx, & Ky \end{pmatrix},$$

where $Kx \cap Ky = 0$, precludes the possibility of its containing a nonzero scalar matrix $\begin{pmatrix} d0 \\ 0d \end{pmatrix}$ with $0 \neq d \in \hat{K}$.

C. Theorem 6 implies that a right quotient-simple ring R which is not an integral domain contains nonzero nilpotent elements. However, such a ring R need not contain non-trivial idempotents even if R contains an identity. Perhaps the simplest example is as follows: Let $S = Q_2$ be the ring of all 2×2 matrices over the rational number field Q, and let R be the subring consisting of all matrices $\begin{pmatrix} a & b \\ c & d \end{pmatrix}$, where b, c are even integers, and a, d are integers which are either both even or both odd. Then $R = (2Z)_2 + Z$ is not an integral domain. However, R is quotient-simple with quotient ring S, with an identity 1, and R does not contain idempotents $\neq 0, 1$.

12. MAXIMAL QUOTIENT RINGS

1. Let R be any right self-injective ring. Then R_R satisfies Baer's condition (§1), so that there is an element $e \in R$ such that $ex = x \ \forall \ x \in R$. Hence R has a left identity e. The set $Q = \{r - re \mid r \in R\}$ is a left ideal of R, and $Q^2 = 0$. Thus, if R is semiprime, then $Q = 0$. Hence any right self-injective semiprime ring has an identity element, a fact used in the proof of the theorem below.

THEOREM 1.1. If S is semiprime and right self-injective, then for any idempotent $e \in S$, eSe is semiprime and right self-injective.

PROOF. Let I be any right ideal of eSe which is nilpotent of index 2. Then

$$(IS)^2 = (IS)(IS) = (IeS)(eIS) = (I(eSe))\, IS \subseteq I^2 S = 0 ,$$

so IS is a nilpotent right ideal of S. Since S is semiprime, $IS = 0$, and $I = 0$, so eSe is semiprime.

Now let I be any right ideal of eSe, let $x = \Sigma_1^n x_i s_i$, $x_i \in I$, $s_i \in S$, $i = 1,\dots,n$ be any element of IS. Let $f \in \mathrm{Hom}_{eSe}(I, eSe)$, and let T denote the set of all elements $\Sigma_1^n f(x_i) r_i \in \Sigma_1^n f(x_i) S$, $r_i \in S$, $i=1,\dots,n$, such that $\Sigma_1^n x_i r_i = 0$. Clearly T is a right ideal of S, and $T \subseteq \Sigma_1^n f(x_i) S \subseteq eS$. Now if $t = \Sigma_1^n f(x_i) r_i \in T$, then

$$te = \left[\Sigma_1^n f(x_i) r_i\right] e = \Sigma_1^n [f(x_i)e] r_i e = \Sigma_1^n f(x_i)(er_i e) =$$

$$\Sigma_1^n f(x_i e r_i e) = f(\Sigma_1^n x_i r_i e) = f(\Sigma_1^n x_i r) = 0.$$

Thus $T^2 = (eT)^2 = 0$, and $T = 0$, since S is semiprime. It follows that $x = \Sigma_1^n x_i s_i = 0$ implies that $\Sigma_1^n f(x_i) s_i = 0$, so that the correspondence

$$f' : x \to \Sigma_1^n f(x_i) s_i ,$$

defined for any $x \in IS$, is an element of $\mathrm{Hom}_S(IS, S)$. Since S_S is injective, and S is semiprime, S has an identity element, so S_S satisfies Baer's condition.

Accordingly, there exists $m \in S$ such that $f'(x) = mx \ \forall \ x \in IS$. In particular, if $x \in I$, then $x = xe$, so that $f'(x) = f(x)e = f(x)$. Thus, $f(x) = (eme)x \ \forall \ x \in I$. Since $eme \in eSe$, this shows that $(eSe)_{eSe}$ satisfies Baer's condition, and eSe is therefore right self-injective.

Below R denotes a ring with $Z_r(R) = 0$, and maximal right quotient ring $\hat{R}$. We recall facts from §8: If I is a right ideal of R, then $\hat{R}$ contains a unique injective hull of I_R, denoted by $\hat{I}_R$, and $\hat{I}$ is the principal right ideal of $\hat{R}$ generated by I. If $C_r(S)$ denotes the lattice of closed right ideals of a ring S, then $C_r(\hat{R})$ consists of principal right ideals, and the contraction map $A \to A \cap R = \bar{A}$ is a lattice isomorphism of $C_r(\hat{R})$ and $C_r(R)$.

THEOREM 1.2. Let R be a semiprime ring with $Z_r(R) = 0$, let I be any right ideal of R, let $\hat{I}$ denote the principal right ideal of $S = \hat{R}$ generated by I, let $\Gamma = \mathrm{Hom}_R(I,I)$, and let $\Delta = \mathrm{Hom}_{\hat{R}}(\hat{I},\hat{I})$. Then $Z_r(\Gamma) = Z_r(\Delta) = 0$, and $\Delta = \hat{\Gamma}$ (= the maximal right quotient ring of Γ.

PROOF. Since S is regular, $\hat{I} = eR$, for some idempotent $e \in S$.

We first show that $\Delta = \mathrm{Hom}_S(\hat{I}_R,\hat{I}_R)$ coincides with $\Omega = \mathrm{Hom}_R(\hat{I},\hat{I})$. Clearly $\Omega \supseteq \Delta$. Conversely if $f \in \Omega$, and if $r \in \hat{I}_R \cap R = \bar{I}$, then $f(r) = f(e)r$. If $x \in \hat{I}_R$, then $x_R = \{t \in R \mid xt \in R\}$ is an essential right ideal of R, and, in fact, $x_R = \{t \in R \mid xt \in \bar{I}\}$. Now if $t \in x_R$, then

$$f(x)t = f(xt) = f(e)xt ,$$

that is, $(f(x) - f(e)x)t = 0$. Thus, $[f(x)-f(e)x]\, x_R = 0$. Since $\hat{I}_R$ has zero singular submodule, we conclude that $f(x) = f(e)x \ \forall \ x \in \hat{I}_R$, and then clearly $f \in \Delta$. This establishes $\Omega = \Delta$.

Since $\hat{I}_R$ is the injective hull of I_R, it follows that each $\gamma \in \Gamma$ has an extension $\hat{\gamma} \in \Delta = \Omega$, and $\hat{\gamma}$ is unique, since $\hat{I}_R$ is rational over I_R. Clearly $\{\hat{\gamma} \in \Delta \mid \gamma \in \Gamma\}$ is a subring of Δ isomorphic to Γ under $\gamma \leftrightarrow \hat{\gamma}$. Henceforth, consider Γ as a subring of Δ.

Now Δ is isomorphic to the ring eSe. If I_L denotes the totality of left multiplications a_L of I by elements $a \in I$,

$$a_L: x \to ax \qquad x \in I,$$

then I_L is a subring of Γ, and the natural isomorphism $\Delta \cong eSe$ maps I_L onto eIe and maps Γ onto a subring Γ_e of eSe. Since $\Gamma_e \supseteq eIe$, in order to show that $(eSe \nabla \Gamma_e)_{\Gamma_e}$, it suffices to show that $(eSe \nabla eIe)_{eIe}$.

Now let $0 \neq \delta \in eSe$. Since $(eS \nabla I)_R$, there exists $r \in R$ such that $0 \neq \delta r \in I$. Since $\delta r = \delta(er)$, it follows that $er \neq 0$. By the same reasoning, since $(eS \nabla I)_R$, there exists $s \in R$, $n \in Z$, such that $u = er(s+n) \in I$, and $w = \delta r(s+n) \neq 0$. Since $\delta r \in I$, it follows that $w \in I$. Since R is semiprime, R is left-faithful, hence $wR \neq 0$ and also $(wR)^2 \neq 0$. Therefore, one can choose $t \in R$ such that $w' = wt$ satisfies $w'' = w'e \neq 0$. Then $w' = \delta u'$, where $u' = ut \in I$ and $w' \in I$, and

$$0 \neq w'' = \delta u'',$$

with $u'' = u'e \in eIe$ and $w'' \in eIe$. Thus, $(eSe \nabla eIe)_{eIe}$ as asserted. Hence eSe is a right quotient ring of Γ_e, and Δ is a right quotient ring of Γ. Now S is regular (hence semiprime), so that $\Delta = eSe$ is right self-injective by Theorem 1.1. Thus, Δ is a maximal right quotient ring of Γ. Since Δ ($\cong eSe$) is regular, $Z_r(\Gamma) = Z_r(\Delta) = 0$.

THEOREM 1.3. Let R be any semiprime ring satisfying $Z_r(R) = 0$, and let e be any idempotent in $S = \hat{R}$. Then:

(1) $eSe = \hat{K}$, where $K = eRe$.

(2) If S is also a left quotient ring of R, then $eSe = \hat{K}$, where $K = eSe \cap R$, and eSe is also a left quotient ring of K.

(3) If e is a primitive idempotent, and if $eSe \cap R \neq 0$, then eSe is the right quotient field of $K = eSe \cap R$.

PROOF. Let $B = (eS \cap R) + (1-e)S \cap R$. The lattice isomorphism $C_r(\hat{R}) \cong C_r(R)$ implies that $[eS \nabla eS \cap R]_R$ (resp. $[(1-e)S \nabla (1-e)S \cap R]_R$), and it follows that

$(S \nabla B)_R$. For each $x \in S$, $x_B = \{b \in B \mid xb \in B\}$ is a right ideal of B. We first show that $(R \nabla x_B)_R$. If $t \in R$, then $(S \nabla B)_R$ implies that

$$t' = \{r \in R \mid tr \in B\}$$

is an essential right ideal of R . If $t \neq 0$, then $Z_r(R) = 0$ implies that $t\,t' \neq 0$. If $xt = 0$, then $0 \subset t\ t' \subseteq x_B \cap tR$. If $xt \neq 0$, then $xt\ t' \neq 0$ and $(S \nabla B)_R$ implies that $xt\ t' \cap B \neq 0$. Then $x_B \cap tR \neq 0$ in this case too, proving that $(R \nabla x_B)_R$.

Now choose $\delta \in eSe$ and $\delta \neq 0$. Then by what we have shown, δ_B is an essential right ideal of R, so $\delta\delta_B \neq 0$ (since $Z(S_R) = Z(R_R) = 0$). By semiprimeness of R, we see that $(\delta\delta_B)^2 \neq 0$, so that $\delta\delta_B e \neq 0$. Hence we can choose $b \in \delta_B \subseteq B$ such that $\delta be \neq 0$. (Note that $eb \in eS \cap R$.)

Case (1). Now $\delta b \in R$, and $b \in R$, and $\delta be = \delta(ebe) = e(\delta b)e$. Thus $0 \neq \delta(ebe)$ eRe with $ebe \in eRe$. This shows that eSe is a right quotient ring of eRe.

Case (2). Since S is a regular ring which is both a right and left quotient ring of R, necessarily $Z_1(R) = 0$. Then ${}_RS$ is a rational extension of ${}_RR$, and, moreover, ${}_R(Se)$ is a rational extension of ${}_R(Se \cap R)$. Now the correspondence $x \to x\delta be\ \forall\, x \in Se$ is an element $f \in Hom_R(Se,Se)$, and $f \neq 0$ since $e(\delta be) = \delta be \neq 0$. It follows that $f(Se \cap R) \neq 0$, that is, that $(Se \cap R)\delta be \neq 0$ and $(Se \cap R)\delta b \neq 0$. By the semiprimeness of R, $[(Se \cap R)\delta b]^2 \neq 0$, and so $\delta b(Se \cap R) \neq 0$. Hence we can choose $u \in Se \cap R$ such that $\delta bu \neq 0$. Since $b, \delta b, u \in R$, then also $\delta bu, bu \in R$, and

$$\delta bu = (e\delta)b(ue) = e(\delta bu)e \in eSe \cap R = K.$$

Since $eb \in eS \cap R$, and $u \in Se \cap R$, then $k = ebu \in K$, so that $0 \neq \delta k = \delta bu \in K$, with $k \in K$. Since δ was an arbitrary nonzero element of eSe, this proves that eSe is a right quotient ring of K.

Case (3). eSe is a division ring and Se is a right vector space over eSe. Since $Se \cap R \neq 0$, and since $0 \neq \delta be \in eSe$, it follows that $(Se \cap R)\delta be \neq 0$, and the rest of the proof proceeds as in the proof of (2).

In all cases we have deduced that eSe is a right quotient ring of K without resource to the fact that the right quotient ring S of R is maximal. Now in Case (2), $Z_1(R) = 0$ is a consequence of the fact that S is a regular ring which is a left quotient ring of R. Hence, by symmetry, we conclude that eSe is a left quotient ring of K in this case.

Since eSe is a regular (von Neumann) ring along with S, and since eSe is right self-injective by Theorem 1.1, we see that $eSe = \hat{K}$ in all cases.

EXAMPLE. $K = eSe \cap R$ need not be semiprime even in the case where S is simple artinian and the classical quotient ring of R. The example is as follows. Let Q be a right Ore domain which is not a left Ore domain. Then Q has a right quotient field D, and there exist elements a, b in Q such that $Qa \cap Qb = 0$. Let e denote the idempotent $e_{11} + e_{12}ab^{-1} + e_{33}$, where e_{ij}, i,j = 1,2,3 are matrix units in the ring $S = D_3$ of all 3 × 3 matrices over D. Note that S is the classical right quotient ring of $R = Q_3$. Now $K = eSe \cap R$ consists of all matrices (q_{ij}) in R such that $e(q_{ij}) = (q_{ij})e = (q_{ij})$. Computation shows that $K = e_{13}Q + e_{33}Q$. Furthermore, K is not semiprime since $e_{13}Q$ is a right ideal of K having square zero.

We next prove a theorem (Theorem 1.6) which shows that, in the example, eSe _is the maximal right quotient ring of_ K. Since K is not prime (nor even semiprime), eSe is therefore _not_ a classical quotient ring of K, even though S is a classical quotient ring of R.

In general R denotes a ring with regular maximal right quotient ring S. The proof of Theorem 1.2 has the following corollary: _If_ R _is semiprime, and if_ I _is a right ideal of_ R, _then_ eSe _is the maximal right quotient ring of_ eIe, _where_ e _is an idempotent in_ S _such that_ eS _is the (unique) injective hull of_ I_R _contained in_ S. _In particular_, $Z_r(eIe) = 0$. Thus, if e is any idempotent in S, and $U = eS \cap R$, then eS is the injective hull in S of U_R, so eSe is the maximal right quotient ring of Ue, and $Z_r(Ue) = 0$.

PROPOSITION 1.4. If R,S,U, and e are as above, and if R is semiprime (resp. prime), then Ue is semiprime (resp. prime).

PROOF. (semiprime). Let I be a left ideal of Ue such that $I^2 = 0$. Then IU is a left ideal of R whose square is zero, so semiprimeness of R implies $IU = 0$. Then, $IUe = 0$, so I is contained in the right singular ideal $Z_r(Ue)$, which is zero by the corollary stated above. Thus, $I = 0$, so R is semiprime.

(prime). Let I, J be left ideals of Ue such that $IJ = 0$. Then IU, JU are right ideals of R and

$$(IU)(JU) = IUeJU \subseteq IJU = 0.$$

Consequently, $IU = 0$, or $JU = 0$, and then by the reasoning above, either $I = 0$, or $J = 0$, so Ue is prime.

PROPOSITION 1.5. A prime ring T is a right quotient ring of any left ideal $L \neq 0$.

PROOF. If $x \in T$, and if $x \neq 0$, then $xL \neq 0$ by primeness of T. Since $xL \subseteq L$, we have $xL \cap L \neq 0$, showing that T is a right quotient ring of L.

THEOREM 1.6. Let R be a prime ring with $Z_r(R) = 0$, and let e be an idempotent in the maximal right quotient ring S of R such that the subring $K = eSe \cap R$ is nonzero. Then eSe is the maximal right quotient ring of K.

PROOF. As stated in the corollary at the beginning, eSe is a right quotient ring of eUe, that is, of Ue, where $U = eS \cap R$. But since R is prime, Proposition 1.4 implies that Ue is prime. Since K is a nonzero left ideal of the prime ring Ue, Ue is a right quotient ring of K by Proposition 1.5. By transitivity of "is a quotient ring of", it follows that eSe is a right quotient ring of K. Since eSe is right self-injective, eSe is necessarily the maximal right quotient ring of K.

2. A REPLACEMENT LEMMA. We first construct an example which shows that (2) and (3) of Theorem 1.3 fail under a weakening of the hypothesis.

Let K be a right Ore domain which is not a left Ore domain, and let k_1, k_2 be nonzero elements of K such that $Kk_1 \cap Kk_2 = 0$.

If D denotes the right quotient division ring of K, then $S = D_2$, the full ring of all 2×2 matrices over D, is the classical, and maximal, right quotient ring of

$R = K_2$. Let $\{e_{ij} \mid i, j = 1, 2\}$ denote matrix units in S, let $a = k_1^{-1}e_{11} + k_2^{-1}e_{12}$, and suppose $b \in S$ is such that $ba \in R$. Then $b = \Sigma_{i,j=1}^{2} c_{ij}e_{ij}$, with $c_{ij} \in D$, $i,j = 1,2$, and

$$ba = c_{11}k_1^{-1}e_{11} + c_{11}k_2^{-1}e_{12} + c_{21}k_1^{-1}e_{21} + c_{21}k_2^{-1}e_{22} .$$

Since $ba \in K_2 = R$, necessarily

$$c_{11}k_1^{-1},\ c_{11}k_2^{-1},\ c_{21}k_1^{-1},\ c_{21}k_2^{-1} \in K ,$$

and then $c_{11}, c_{21} \in Kk_1 \cap Kk_2$. Since $Kk_1 \cap Kk_2 = 0$, $c_{11} = c_{22} = 0$, so necessarily $ba = 0$. This shows that $Sa \cap R = 0$. Now $e = k_1a = e_{11}+k_1k_2^{-1}e_{12}$ belongs to Sa, and $e = e^2$. It is easy to see that eSe is a division ring (or equivalently, that Se is a minimal left ideal of S), while $eSe \cap R = 0$. In particular, eSe is not a quotient ring of $eSe \cap R$.

In view of this example, the following "replacement" lemma is of interest.

LEMMA 2.1. Let R be a prime ring, and let g be a primitive idempotent in a right quotient ring S of R. Then there exists an idempotent $e \in S$ such that $eS = gS$ and such that $eSe \cap R \neq 0$.

PROOF. We first note the trivial fact that S is prime (along with R). Select $a \in S$ such that $0 = ga \in R$. Then $I = R \cap Sga \neq 0$; also $U = gS \cap R \neq 0$. Since R is prime, $IRU \neq 0$, so there exists $b \in R$ such that $Ibg \neq 0$. Then $0 \neq Ib \subseteq Sgab$, and so $R \cap Sgab \neq 0$. Since Sg is a minimal left ideal of S, so is $H = Sgab$. Hence there exists a primitive idempotent $h \in S$ such that $H = Sh$. But, since $Shg \supseteq Ibg \neq 0$, $hg \neq 0$, and $ShgS \neq 0$. Then, since S is prime, it is well-known that there exists an idempotent $e \in S$ such that $H = Se$ and $gS = eS$. Then, by primeness of R, $0 \neq (R \cap eS)(R \cap Se) \subseteq R \cap eSe$.

3. LEFT AND RIGHT QUOTIENT RINGS. In view of Theorem 1.3, it is of interest to consider conditions which imply that the maximal right quotient ring is also a left quotient ring. The general question has been extensively treated by Utumi [1]. Here we shall develop different conditions for prime rings containing a minimal closed right ideal. Below, a ring Γ is _left Ore_ in case $\Gamma x \cap \Gamma y \neq 0$ for each pair of nonzero

$x, y \in \Gamma$.

THEOREM 3.1. Let R be a prime ring with zero right singular ideal, and containing a minimal closed right ideal U. Let S be the maximal right quotient ring of R. Then S is also a left quotient ring of R if and only if the following two conditions are satisfied:

(1) $\Gamma = \mathrm{Hom}_R(U,U)$ is a left Ore ring;

(2) $\Delta U = US$, where $\Delta = \mathrm{Hom}_S(US,US)$.

PROOF. The proof will show that the theorem holds even if U is assumed to be any nonzero right ideal contained in a minimal closed right ideal $\bar{U}$.

Since S is a regular ring, the lattice isomorphism $C_r(S) \cong C_r(R)$ shows that $\bar{U}S$ is a minimal right ideal of S. Hence US $(= \bar{U}S)$ is a minimal right ideal of S, so there exists $e = e^2 \in S$ such that $US = eS$, and then $D = eSe$ is a division ring. By Theorem 1.2, Δ is the maximal right quotient ring of Γ (making Γ a <u>right</u> Ore domain). The canonical isomorphism $\Delta \cong D = eSe$ maps Γ isomorphically onto a subring Γ^* of Δ , and $\Gamma^* \supseteq K = eSe \cap R$.

Now assume that S is also a left quotient ring of R. Then (2) of Theorem 1.3, states that $K = eSe \cap R$ is a left and right Ore domain whose quotient division ring is $D = eSe$. Since $\Gamma^* \supseteq K$, this implies that Γ^*, hence Γ , is a left Ore domain, proving (1). If $0 \neq x \in S$, then $L_x = Sx \cap R \neq 0$. If $0 \neq x \in eS = US$, then $L_x \cap eS = Dx = \Delta x$. Since primeness of R implies that $UL_x \neq 0$, we can choose $0 \neq u \in UL_x$. Since $UL_x \subseteq L_x \cap U \subseteq L_x \cap eS = \Delta x$, we have $0 \neq u \in \Delta x$. Since eS is a left vector space over Δ, it follows that $\Delta u = \Delta x$, establishing that $eS \subseteq \Delta U$ and $eS = \Delta U$. This proves (2).

Conversely, assume (1) and (2). If $0 \neq a \in S$, then $(eS)a = (\Delta U)a = \Delta(Ua) \neq 0$, so $Ua \neq 0$. Pick $u \in U$ such that $w = ua \neq 0$. Since $w \in eS = \Delta U$, there exist $\delta_1,\dots,\delta_n \in \Delta$, $u_1,\dots,u_n \in U$ such that $w = \Sigma_1^n \delta_i u_i$. Since Γ is left Ore (with quotient field Δ), there exists $0 \neq \gamma \in \Gamma$ such that $0 \neq \gamma\delta_i = \gamma_i \in \Gamma$, $i=1,\dots,n$. Since $\Gamma U \subseteq U$, we see that

$$0 \neq \gamma w = (\gamma u)a = \sum_{i=1}^{n} \gamma_i u_i \in Ua \cap U.$$

Thus, $Ra \cap R \neq 0 \; \forall \; 0 \neq a \in S$, and S is therefore a left quotient ring of R.

The theorem generalizes Goldie's theorem for classical quotient rings (Goldie [2, p.218]).

13. QUOTIENT RINGS AND DIRECT PRODUCTS OF FULL LINEAR RINGS

1. **THE ENDOMORPHISM RING OF AN INJECTIVE MODULE.** Let R be a ring (in which an identity element is not assumed), let ${}_RQ$ be an injective R-module, and let $S = \mathrm{Hom}_R(Q,Q)$. A result of Utumi [2, p.19, Lemma 8] states that the Jacobson radical $J(S)$ of S can be described by

(1) $J(S) = \{\alpha \in S \mid Q \nabla \ker(\alpha)\}$,

that is, $J(S)$ is the set of $\alpha \in S$ such that $\ker(\alpha)$ is an essential R-submodule of Q. Furthermore (loc.cit.):

(2) $S - J(S)$ is a (von Neumann) regular ring.

Thus, $J(S) = 0$ if and only if S is a regular ring.

Throughout this section we shall assume that Q is an injective left R-module such that: (a) $S = \mathrm{Hom}_R(Q,Q)$ is a regular ring; and (b) every direct summand of ${}_RQ$ contains a minimal direct summand (m.d.s.). (See (2.5) below for a sufficient condition for (a).)

THEOREM 1.1. An R-submodule Q' of Q is a direct summand of Q if and only if it is the annihilator of a finite subset of S.

PROOF. If Q' is a direct summand of Q, then $Q' = Qe$, where e is an idempotent of S, in which case $Q' = \{1-e\}^Q$. Conversely, let $Q' = \{\alpha_1,\dots,\alpha_r\}^Q$, with $\alpha_1,\dots,\alpha_r \in S$. Then, since S is regular, there exists an idempotent $e \in S$ such that $\alpha_1 S + \dots + \alpha_r S = eS$ (von Neumann, On regular rings, Proc. Nat. Acad. Sci. (U.S.A.) 22 (1936) 707-713), in which case $Q' = Q(1-e)$, a direct summand of Q.

THEOREM 1.2. If Q' is a direct summand of Q, and if Q_o is a m.d.s. of Q, then either $Q_o \cap Q' = 0$ or $Q_o \subseteq Q'$.

PROOF. We have immediately from 1.1 that $Q_o \cap Q'$ is a direct summand of Q. Since Q_o is a m.d.s., 1.2 follows.

THEOREM 1.3. If Q_o is a m.d.s. of Q and $\alpha \in S$, then either $Q_o\alpha = 0$ or $Q_o\alpha$ is a m.d.s. of Q isomorphic to Q_o.

PROOF. $Ker(\alpha)$ is a direct summand of Q by 1.1. Hence, by 1.2, either $Q_o \subseteq ker(\alpha)$ or $Q_o \cap ker(\alpha) = 0$. In the former case $Q_o\alpha = 0$; in the latter, $Q_o\alpha \cong Q_o$, so $Q_o\alpha$ is injective along with Q_o . This shows that $Q_o\alpha$ is a direct summand, and $Q_o\alpha \cong Q_o$ yields that Q_o is a m.d.s.

Now, by Zorn's lemma, we select a family $\{Q_i\}$ of m.d.s.'s of Q which is maximal with respect to the property that the sum $C = \Sigma_i Q_i$ is direct. In the next few statements we focus our attention on C.

THEOREM 1.4. IF Q_o is a m.d.s. of Q, then there exist $i_1,\ldots,i_r$ such that $Q_o \subseteq Q_{i_1} \oplus \ldots \oplus Q_{i_r}$ and $Q_{i_k} \cong Q_o \ \forall\, k \leq r$.

PROOF. If $Q_o \cap C = 0$, then $\{Q_i, Q_o\}$ is a family of m.d.s.'s of Q whose sum is direct, contradicting the maximality of the family $\{Q_i\}$ with respect to this property. Hence, there exist $i_1,\ldots,i_r$ such that $Q_o \cap (Q_{i_1} \oplus \ldots \oplus Q_{i_r}) \neq 0$. Since $Q_{i_1} \oplus \ldots \oplus Q_{i_r}$ is injective, and therefore a direct summand of Q, we have from 1.2 that $Q_o \subseteq Q_{i_1} \oplus \ldots \oplus Q_{i_r}$. We may assume that Q_o has nonzero projection on each Q_{i_k} , $k \leq r$. This projection is a homomorphism of Q_o into Q which can be induced by an element $\alpha_k \in S$. Since $0 \neq Q_o\alpha_k \subseteq Q_{i_k}$, it follows from 1.3, that $Q_o\alpha_k \cong Q_o$, and that $Q_o\alpha_k$ is a m.d.s. of Q_{i_k} . Since Q_{i_k} is likewise a m.d.s. of Q containing $Q_o\alpha_k$, we see that $Q_{i_k} = Q_o\alpha_k \cong Q_o \ \forall\, k \leq r$, completing the proof.

THEOREM 1.5. ${}_RC$ is an essential submodule of ${}_RQ$. Hence, $Q \cong E({}_RC)$.

PROOF. Let $Q' \neq 0$ be a submodule of ${}_RQ$. Since Q is injective, we can assume that $Q \supseteq E(Q')$, the injective hull of Q'. Then $E(Q')$ is a direct summand of Q, hence contains a m.d.s. Q_o . But $Q_o \subseteq C$ by 1.4, and so $C \cap Q' \neq 0$. Thus, C is an essential submodule of Q, and $Q = E({}_RC)$ follows.

THEOREM 1.6. C is the R-submodule of Q generated by all m.d.s.'s of Q, and is an (R,S)-bisubmodule of Q.

PROOF. C is by construction generated by m.d.s.'s of Q, and by 1.4, C contains each m.d.s. of Q, proving the first part. On the other hand, if Q_o is a m.d.s. of Q and if $\alpha \in S$, then it follows from 1.3 and 1.4 that $Q_o\alpha \subseteq C$, and hence $C\alpha \subseteq C$. Thus C is an (R,S)-bisubmodule.

THEOREM 1.7. The natural mapping $S \to \mathrm{Hom}_R(C,C)$, defined by restriction, is a ring isomorphism.

PROOF. The mapping is obviously a ring homomorphism. If $\alpha \in S$ and $\alpha \to 0$ then $C\alpha = 0$. Since $Q \nabla C$, this implies that $\alpha \in J(S)$, so $\alpha = 0$. Thus, the map is a ring monomorphism; it is a ring epimorphism since ${}_RQ$ is injective.

We now summarize the results so far obtained.

THEOREM 1.8. Let Q be an injective left R-module satisfying the conditions: (a) $S = \mathrm{Hom}_R(Q,Q)$ is a regular ring; (b) every direct summand of ${}_RQ$ contains a m.d.s. Let C be the submodule generated by all m.d.s.'s of Q. Then C is a fully invariant submodule of Q, and is a direct sum of m.d.s.'s. Furthermore, the natural mapping $\mathrm{Hom}_R(Q,Q) \to \mathrm{Hom}_R(C,C)$ is a ring isomorphism.

THEOREM 1.9. If Q_o is a m.d.s. of Q, then $\mathrm{Hom}_R(Q_o,Q_o)$ is a sfield.

PROOF. Let $S_o = \mathrm{Hom}_R(Q_o,Q_o)$, and write $Q_o = Qe$, where $e = e^2 \in S$. Then $S_o = eSe$ is a regular ring along with S. Since Q_o is indecomposable, S_o has no non-trivial idempotents, hence S_o must be a sfield.

Now let T be the set of all isomorphism classes of m.d.s.'s of Q, and for each $\tau \in T$ let $Q^{(\tau)}$ be a representative from the class τ. Set $A = \Sigma_{\tau\in T} \oplus Q^{(\tau)}$ and $K = \mathrm{Hom}_R(A,A)$. Then K is a ring, and A is an (R,K)-bimodule. Also let $V = \mathrm{Hom}_R(A,C)$; then V is a left K-module. If $x \in A$, $y \in V$, we denote the value of y on x by xy. Let $\sigma: A\otimes_K V \to C$ denote the valuation map, i.e., $\sigma(x\otimes y) = xy \quad \forall\, x \in A,\ y \in V$.

THEOREM 1.10. The mapping $\sigma: A\otimes_K V \to C$ is a left R-module isomorphism.

PROOF. σ is evidently a left R-homomorphism. If $z \in Q_i$, then there exists $\tau \in T$ such that $Q^{(\tau)} \cong Q_i$, in which case we can select $y \in V$ such that $Q^{(\tau)}y = Q_i$. If $x \in Q^{(\tau)}$ and $xy = z$, then $\sigma(x\otimes y) = xy = z$. Since $C = \Sigma_i \oplus Q_i$, it follows that σ is onto.

Now suppose $\sigma(u) = 0$, where

$$u = x_1 \otimes y_1 + \dots + x_n \otimes y_n \quad A\otimes_K V .$$

We may assume that $x_k \in Q^{(\tau_k)}$ for some $\tau_1,\dots,\tau_n \in T$. Then there exists an idempotent $e_k \in K$ such that $Q^{(\tau_k)}(1-e_k) = 0$ and $Q^{(\tau)}e_k = 0$ if $\tau \neq \tau_k$, in which case $x_k \otimes y_k = x_k \otimes e_k y_k$. Replacing y_k by $e_k y_k$, we may assume that $Q^{(\tau)}y_k = 0 \ \forall \ \tau \neq \tau_k$.

Next we note the critical fact that 1.4 implies the existence of $i_1,\dots,i_r$ (depending on k) such that

$$Ay_k = Q^{(\tau_k)}y_k \subseteq Q_{i_1} \oplus \dots \oplus Q_{i_r} .$$

Then, $y_k = y_{k_1} + \dots + y_{k_r}$, where $y_{k_t} \in V$ is such that $Q^{(\tau_k)}y_{k_t} \subseteq Q_{i_t}$, and $Q^{(\tau)}y_{k_t} = 0$ if $\tau \neq \tau_k$ $(t \leq r)$. Hence, expanding the expression for u, using the associative law, we may assume that $Q^{(\tau_k)}y_k \subseteq Q_{i_k}$ for some i_k; and, of course, $Q^{(\tau_k)} \cong Q_{i_k}$ by 1.4.

Since $Q^{(\tau_k)} = Q_{i_k}$, it follows from 1.4 that $\mathrm{Hom}_R(Q^{(\tau_k)}, Q_{i_k})$ is a one-dimensional left vector space over the sfield $\mathrm{Hom}_R(Q^{(\tau_k)}, Q^{(\tau_k)})$, a direct summand of K. Hence, multiplying each x_k and y_k, if necessary, by suitable elements of K, we may ensure that $y_\mu = y_\nu$ whenever $i_\mu = i_\nu$. Collecting terms in the expression for u, we may then assume that $i_1,\dots,i_r$ are all distinct and each $y_{i_k} \neq 0$, in which case y_{i_k} is a monomorphism by 1.3. Then $0 = \sigma(u) = x_1y_1 + \dots + x_ny_n$, and so $x_ky_k = 0 \ \forall \ k \leq n$, since $x_ky_k \in Q_{i_k}$ and $i_1,\dots,i_n$ are all distinct. But each y_k is a monomorphism, so each $x_k = 0$ and $u = 0$. Thus σ is a monomorphism, and therefore an isomorphism, completing the proof.

THEOREM 1.11. $S = \mathrm{Hom}_K(V,V)$, a direct product of full right linear rings.

PROOF. We have seen in 1.8 that $S = \mathrm{Hom}_R(C,C)$. Viewing C is an (R,S)-bimodule, the fact that $V = \mathrm{Hom}_R(A,C)$ defines in the usual way a right S-module structure on V such that V is a (K,S)-bimodule. Let φ denote the natural map $\varphi: S \to S' = \mathrm{Hom}_K(V,V)$.

Now, viewing V as a (K,S)-bimodule, we see that the isomorphism $\sigma: A \otimes_K V \to C$ defines in the obvious way a right S'-module structure on C such that C becomes an (R,S')-bimodule. Hence we obtain a ring homomorphism $\varphi': S' \to S$. It is easily checked that $\varphi\varphi'$ and $\varphi'\varphi$ are the identity maps on S' and S respectively. Thus, φ and φ' are isomorphisms, and $S \cong S' = \mathrm{Hom}_K(V,V)$.

By 1.3, $\mathrm{Hom}_R(Q^{(\tau)},Q^{(\tau')}) = 0$ if $\tau \neq \tau'$. From this we immediately see that $K \cong \Pi_\tau K^{(\tau)}$, where, by 1.9, $K^{(\tau)} = \mathrm{Hom}_R(Q^{(\tau)},Q^{(\tau)})$ is a sfield $\forall \tau \in T$. Then there exists a K-isomorphism $V \cong \Pi_\tau V^{(\tau)}$, where $V^{(\tau)}$ is a left $K^{(\tau)}$-space. It follows that

$$S \cong \mathrm{Hom}_K(V,V) \cong \Pi_\tau \mathrm{Hom}_{K(\tau)}(V^{(\tau)},V^{(\tau)}) ,$$

a direct product of full right linear rings. This completes the proof.

To summarize:

THEOREM 1.12. Let Q be an injective left R-module satisfying the conditions: (a) $S = \mathrm{Hom}_R(Q,Q)$ is a regular ring; (b) every direct summand of ${}_RQ$ contains a minimal direct summand. Then, S is a direct product of full right linear rings.

We next prove two statements preparatory to a converse of Theorem 1.12.

THEOREM 1.13. Let ${}_RM$ be any module, and $T = \mathrm{Hom}_R(M,M)$. Then the following two statements are equivalent:

(1) Each nonzero direct summand of ${}_RM$ contains a m.d.s.

(2) If $\mathcal{F}$ denotes the family of the left ideals of T of the form Te, where e is idempotent of T, then each nonzero member of $\mathcal{F}$ contains a minimal nonzero member of $\mathcal{F}$.

PROOF. Let $\mathcal{F}^r$ denote the family of right ideals of T derived from $\mathcal{F}$ by left-right symmetry. Then $Te \to (1-e)T$ is a 1-1 order-inverting correspondence between $\mathcal{F}$ and $\mathcal{F}^r$, and $(1-e)T \to M(1-e)$ is 1-1 order-inverting between $\mathcal{F}^r$ and the set p of direct summands of M. Hence $Te \to M(1-e)$ is 1-1 and order-preserving between $\mathcal{F}$ and p. The statement now follows.

If T is regular, then each principal left ideal of T is generated by an idempotent (von Neumann, On regular rings, Proc. Nat. Acad. Sci. (U.S.A.) 22 (1936) 707-713). This, together with the fact that a minimal principal left ideal is actually a minimal left ideal, yields the next statement.

THEOREM 1.14. Let M be a left R-module such that $T = \mathrm{Hom}_R(M,M)$ is a regular ring. Then the following conditions are equivalent:

(1) Each nonzero direct summand of $_RM$ contains a m.d.s.

(2) Each nonzero left ideal of T contains a minimal left ideal.

The last statement of this section is a converse to Theorem 1.12.

THEOREM 1.15. Let M be a left R-module, and assume that $T = \mathrm{Hom}_R(M,M)$ is isomorphic to a direct product of full right linear rings. Then each nonzero direct summand of M contains a m.d.s.

PROOF. Write $T = \Pi_i T_i$, where T_i is a full right linear ring. Let S_i (= the socle of T_i) denote the sum of all minimal left ideals of T_i. Then, S_i is a completely reducible left T_i-module, and a nonzero left ideal K of T_i has nonzero intersection with S_i. Thus K contains a minimal left ideal of T_i. If follows that each nonzero left ideal of T contains a minimal left ideal. Since T, a direct product of regular rings, is regular, we can apply 1.14, to complete the proof.

2. RATIONALLY CLOSED SUBMODULES. Most of the results in this section on rationally closed submodules are essentially known, yet they do not appear in the literature in a form which we require (mainly because we start with a different definition).

If P is a submodule of a left R-module M, and if $x \in M$, then $(P:x) = \{r \in R \mid rx \in P\}$ is a left ideal of R. If X is a subset of R, then $X^M = \{m \in M \mid Xm = 0\}$.

If $X \subseteq M$, then $X^R = \{r \in R \mid rX \subseteq R\}$ is a left ideal of R.

DEFINITION. A submodule P of M is said to be <u>rationally closed</u> (r.c.) <u>in</u> M in case the following condition holds: If $x \in M$, and if $(P:x)^M = 0$, then $x \in P$. We let $C(M)$ denote totality of closed submodules of M; $C(M)$ is nonempty, since 0, $M \in C(M)$.

The notion just defined is related to a concept of Findlay and Lambek [1]. (Cf. Lambek [1].)

THEOREM 2.1. $C(M)$ is a complete lattice.

PROOF. Let $\{P_i\}$ be an arbitrary family of r.c. submodules. Clearly $\cap_i(P_i:x) = (\cap_i P_i : x)$ $\forall\, x \in M$, so that

$$\Sigma_i(P_i : x)^M \subseteq (\cap_i P_i : x)^M \qquad \forall\, x \in M .$$

Thus, $(\cap_i P_i : x)^M = 0$ implies that $(P_i : x)^M = 0\ \forall\, i$. Since $P_i \in C(M)\ \forall\, i$, it follows that $x \in \cap_i P_i$ so that $\cap_i P_i$ is r.c. Thus, each submodule N of M has a unique minimal r.c. cover N', namely, N' = the intersection of all r.c. submodules of M containing N. Defining $P \vee K = (P + K)'\ \forall\, P, K \in C(M)$, $C(M)$ is obviously a complete lattice under $\vee$ and $\cap$.

We first apply this concept for the case $M = R$. In this case, let $X^r = \{a \in R \mid Xa = 0\}$, $X^l = \{a \in R \mid aX = 0\}$, for $X \subseteq R$. Then a left ideal P is r.c. in R (viewing P as a submodule of $_RR$) if $(P:x)^r = 0$ implies that $x \in P$.

THEOREM 2.2. Any annihilator left ideal of R is r.c.

PROOF. Let I be a right ideal of R, and let $x \in R$, $x \notin I^l$. Then $xI \neq 0$, but $xI \subseteq I^l : x)^r$. Thus $(I^l : x)^r \neq 0$, showing that the annihilator left ideal $P = I$ is r.c.

If $P \supseteq N$ are submodules of M, then $P \triangledown N$ signifies that P is an <u>essential extension of</u> N, that is, that $K \cap N \neq 0$ for each submodule $K \neq 0$ of P. N is <u>closed in</u> M in case $P \triangledown N \to P = N$ for all submodules P of M.

THEOREM 2.3. Each closed submodule of M is r.c.

PROOF. Let Q be the injective hull of the left R-module M, and let Q' be an injective hull in Q of a closed submodule N of M. Then $(Q' \cap M) \triangledown N$, so $Q' \cap M = N$, since N is closed. Since Q' is injective, $Q = Q' \oplus Q''$ for some submodule Q" of Q. If $N = M$, then N is r.c. If $N \neq M$, and if $x \in M$, $x \notin N$, then $x \notin Q'$. Writing $x = x' + x''$, with $x' \in Q'$, $x'' \in Q''$, it follows that $x'' \neq 0$. Clearly $(N : x) \subseteq (x'')^R$, so that

$$x'' \in (x'')^{RM} \subseteq (N : x)^M ,$$

showing that $(N : x)^M \neq 0$. Thus N is r.c. in M.

The following result can be established under a more general hypothesis, e.g., assuming only that $\mathrm{Hom}_R(Q,Q)$ is regular, but for brevity we restrict ourselves to the assumption $Z(_RQ) = 0$. (In this case the result is due to R.E. Johnson). We let $E(_RM)$

denote the injective hull of a left R-module ${}_RM$. When possible, we suppress R and write E(M), Z(M), C(M), etc.

THEOREM 2.4. Let ${}_RM$ be a module, let $Q = E(M)$, and assume that $Z(M) = 0$. Then $C(Q)$ consists of the direct summands of Q, and $C(Q)$ is isomorphic to $C(M)$ under the contraction map $Q' \to Q' \cap M$.

PROOF. Since $Z(M) = Z(Q) \cap M = 0$, and since $Q \triangledown M$, it follows that $Z(Q) = 0$. By 2.3, if N is a closed submodule of M, then $N \in C(M)$. Conversely, if $N \in C(M)$, and if P is a submodule of M such that $P \triangledown N$, then $(N : x)$ is an essential left ideal of R, for each $x \in P$. Since $Z(M) = 0$, it follows that $(N : x)^M = 0 \ \forall\, x \in P$. Since N is r.c. in M, this means that $x \in N \ \forall\, x \in P$, so $N = P$, and N is closed in M.

By the result of Eckmann and Schopf [1] (for the non-unital case see Faith and Utumi [1]), the closed submodules of Q are simply the direct summands of Q. This proves the first statement. The lattice isomorphism $C(Q) = C(M)$ follows from a more general result of R.E. Johnson [4, p.1387, Theorem 2.5]*). It follows essentially from the observation, due to Johnson, that each submodule N of Q has a unique maximal essential extension N' in Q (and N' is then a direct summand of Q).

COROLLARY. Under the assumptions of 2.4, $S = \mathrm{Hom}_R(Q,Q)$ is a regular ring.

PROOF. Let $\alpha \in S$, let $K = \ker(\alpha)$, and let P be a submodule of Q such that $P \triangledown K$. Then, if $x \in P$, $(K : x)$ is an essential left ideal of R, so $(K : x)^Q = 0$. Thus, if $x \notin K$, that is, if $x\alpha \neq 0$, then $(K : x)(x\alpha) \neq 0$. But

$$(K : x)(x\alpha) = ((K : x)x)\alpha \subseteq K\alpha = 0 \ ,$$

so $(K : x)x\alpha = 0$, a contradiction showing that $P = K$. Since K is therefore a closed submodule of Q, by 2.4, K is a direct summand of Q. Thus, $K = \ker(\alpha)$ is an essential submodule if and only if $K = Q$, that is, if and only if $\alpha = 0$. By a result of Utumi cited at the beginning of the next section, S is a regular ring.

*) This is given in section 7, Corollary 8, on p. 61 of these Lectures.

3. **MAXIMAL QUOTIENT RINGS.** Let R be a ring in which an identity element is not assumed. A left quotient ring of R is an overring S containing an identity element such that for each $a \in S$ there corresponds $r \in R$ such that $ra \in R$ and $ra \neq 0$. Johnson's [3, Theorem 3] states that R possesses a left quotient ring which is a regular ring if and only if $Z_1(R) = 0$. In this case R possesses a unique (up to isomorphism over R) maximal left quotient ring $\hat{R}$, and $\hat{R}$ is regular and left self-injective. If Q denotes the injective hull of ${}_R R$, Johnson and Wong [2, p.172, Theorem 7]*) characterize $\hat{R}$ as follows: $\hat{R} \cong \mathrm{Hom}_R(Q,Q)$. It follows from rather elementary considerations that $\hat{R}$ and Q are isomorphic left R-modules. In other words, when $Z_1(R) = 0$, then the injective hull Q of ${}_R R$ is a left self-injective regular ring containing R as a subring (in fact, ring multiplication in Q induces the R-module operation defined in Q).

In this section we study rings R such that R is isomorphic to a direct product of full linear rings.

Recall that the left singular ideal $Z_1(R)$ of a ring R is defined by $Z_1(R) = \{x \in R \mid x^l$ is an essential left ideal of $R\}$. Here $X^l = \{a \in R \mid aX = 0\}$ (and $X^r = \{a \in R \mid Xa = 0\}$) for a subset X of R. We first deduce a class of conditions sufficient to insure the vanishing of $Z_1(R)$. An annihilator left ideal is a left ideal of the form X^l for some $X \subseteq R$.

THEOREM 3.1. Let R be a semiprime ring, and let $\mathfrak{F}$ be a nonempty family of left ideals of R subject to the following conditions:

(a) Every annihilator left ideal of R belongs to $\mathfrak{F}$.

(b) Each nonzero member of $\mathfrak{F}$ contains a minimal nonzero member of $\mathfrak{F}$.

(c) If $I, J \in \mathfrak{F}$, then $I \cap J \in \mathfrak{F}$.

Then $Z_1(R) = 0$.

PROOF. We can assume that $R \neq 0$, so that the set $M(\mathfrak{F})$ of minimal nonzero members of $\mathfrak{F}$ is nonempty by (b). Let J be the left ideal of R generated by $M(\mathfrak{F})$. If $I \in M(\mathfrak{F})$, and if $x \in Z_1(R)$, then $x^l \cap I \neq 0$. But $x^l \cap I \in \mathfrak{F}$ by (a) and (c), and

---

*) See p.69 of these Lectures, Theorem 1 and 2.

$I \in M(\mathcal{J})$, so it follows that $x^1 \supseteq I$. Since $Ix = 0$ for all $x \in Z_1(R)$ and all $I \in M(\mathcal{J})$, it follows that $ZJ = 0$, $Z = Z_1(R)$. Then semiprimeness of R implies that $JZ = 0$. If $Z \neq 0$, then $J^1 \neq 0$, and therefore J^1 contains an element $I \in M(\mathcal{J})$, by (a) and (b). Then $I \subseteq J$ and $IJ = 0$, so $I^2 = 0$. This contradicts the facts that R is semiprime and $I \neq 0$. This $Z = Z_1(R) = 0$.

THEOREM 3.2. Let R be a semiprime ring such that $Z_1(R) = 0$, and let $S = \hat{R}$. Then S is prime if and only if R is prime.

PROOF. If R is prime, then trivially S is. Conversely, let I, J be left ideals of R such that $IJ = 0$. Then JSI is a left R-submodule of S, and $R \cap JSI$ is therefore a left ideal of R. Clearly $(R \cap JSI)^2 = 0$, so semiprimeness of R yields $R \cap JSI = 0$. Since S is a left quotient ring of R, we deduce that $JSI = 0$. Then primeness of S implies $I = 0$ or $J = 0$. Thus R is prime, completing the proof.

In the remainder of this section we will consider a special type of subdirect sum of rings. Let R be a subdirect sum of a family $\{R_i\}$ of rings (that is, $R \subseteq \Pi_i R_i$ and the projection $R \to R_i$ is onto $\forall i$ $i \in I$.) Let J be the left ideal in R generated by all the ideals $R \cap R_i$. The subdirect sum will be called <u>essential irredundant</u> if J is an essential left R-submodule of $\Pi_i R_i$. (It is easy to see that an essential irredundant subdirect sum is irredundant in the sense of L. Levy, [Unique direct sums of prime rings, Trans. Amer. Math. Soc. 106 (1963) 64-76]).

THEOREM 3.3. Let R be a ring such that $Z_1(R) = 0$, and let e be a central idempotent in $S = \hat{R}$. Then $Se = \hat{K}$, where $K = Re$. If R is semiprime, so is K.

PROOF. Since $R \cap Se$ is an essential left R-submodule of Se, it is easy to see that Se is an essential left K-submodule of $K = Re$. Furthermore, the ring Se is left self-injective, since it is a direct summand of the ring S and S is left self-injective. Since $Se = eSe$ is a regular ring, it follows that $Z_1(K) = 0$, and that $Se = \hat{K}$.

Suppose that R is semiprime, and that N is a nilpotent left ideal of K. Then $N \cap R$ is a nilpotent left ideal of R, so $N \cap R = 0$. Since N is an R-submodule of S, it follows that $N = 0$, and K is semiprime.

THEOREM 3.4. Let R be a semiprime ring with the property that every r.c. left ideal of R contains a minimal r.c. left ideal. Then: (1) $Z_1(R) = 0$, and $\hat{R} = \Pi_i S_i$, where S_i is a full right linear ring $\forall i$; (2) If e_i is the identity element of S_i, and if $R_i = Re_i$, then R_i is a prime ring, and $S_i = \hat{R}_i$; (3) the mapping $R \to \Pi_i R_i$ defined by $a \to (ae_i)$ (where $a \in R$) gives a representation of R as an essential, irredundant subdirect sum of the prime rings $\{R_i\}$.

PROOF. (1) Let $\mathcal{F}$ be the family of all r.c. left ideals of R. Then (2.2) shows that $\mathcal{F}$ satisfies (a) of (3.1), (b) of (3.1) is satisfied by hypothesis, and (c) of (3.1) follows from (2.1). Thus, (3.1) implies that $Z_1(R) = 0$. If Q is the injective hull of ${}_R R$, then $\hat{R} \cong \mathrm{Hom}_R(Q,Q)$ by the result of Johnson and Wong stated at the beginning of this section. We can apply (2.4) with $M = R$, since $Z(M) = Z_1(R) = 0$. Thus, since each r.c. left ideal of R contains a minimal r.c. left ideal, it follows that each nonzero direct summand of Q contains a m.d.s. Then, Theorem 1.12 applies, so $\hat{R} \cong \mathrm{Hom}_R(Q,Q)$ has the desired structure.

(2) R_i is semiprime and $S_i = \hat{R}_i$ by (3.3); R_i is prime by (3.2).

(3) Now define a mapping $\varphi: R \to \Pi_i R_i$ by $\varphi(a) = ae_i$, where $a \in R$. Clearly φ defines a representation of R as a subdirect sum of the rings $\{R_i\}$, and the composition of φ with the obvious mapping of $\Pi_i R_i$ into $S = \hat{R}$ is simply the inclusion mapping of R into S. Hence, in order to show that the subdirect sum is essential irredundant, we need only show that J, the left ideal of R generated by all $R \cap R_i$, is an essential left R-submodule of S. (Hence we are identifying R with its image in S.)

Let $0 \neq x \in S$. Then $e_i x \neq 0$ for some i. Let $I = \{a \in R \mid ae_i \in R\}$ and $I' = \{a \in R \mid ae_i x \in R\}$. Then I and I' are essential left ideals of R, since R is an essential left R-submodule of $S = \hat{R}$. Since $Z({}_R S) = 0$, $R \cap (e_i x)^1$ is not an essential left ideal of R. Hence there exists $a \in I \cap I'$ such that $ae_i x \neq 0$. Then $b = ae_i \in R$ and $bx = bxe_i$ is a nonzero element of $R \cap R_i$. It follows that J is an essential left R-submodule of S, completing the proof.

If R is a prime ring containing a minimal nonzero r.c. left ideal, then, by the proof of (3.4), $Z_1(R) = 0$, and R is a full right linear ring. This case is due to

Utumi [1, p.11] (cf. Lambek [1, p.401]).*)

We now prove a converse to Theorem 3.4.

THEOREM 3.5. Let $\{R_i\}$ be a family of prime rings such that $Z_1(R_i) = 0 \ \forall i$ and assume that $S_i = \hat{R}_i$ is a full right linear ring. Let R be an essential irredundant subdirect sum of the family $\{R_i\}$. Let $S = \Pi_i S_i$. Then: (1) $S \cong \hat{R}$, the inclusion map of R into S being the composition of the inclusion mapping of R into $\Pi_i R_i$ with the obvious mapping of $\Pi_i R_i$ into S; (2) R is semiprime and each nonzero r.c. left ideal of R contains a minimal nonzero r.c. ideal.

PROOF. As before, let J be the left ideal of R generated by all $R \cap R_i$. Then, by hypothesis, J is an essential left R-submodule of $\Pi_i R_i$. Observe that $R \cap R_i = J \cap R_i \neq 0 \ \forall i$.

Embedding $\Pi_i R_i$ into S in the manner described above, we may view R as a subring of S. S, a direct product of left self-injective regular rings S_i, is itself left self-injective and regular. Hence, in order to show that $S \cong \hat{R}$, we need only show that $(S \nabla R)$. Let $0 \neq x \in S$. Then, for some i, $e_i x \neq 0$, where e_i is the identity of S_i. Then, $e_i x \in S_i$, so $R_i x \cap R_i = R_i e_i x \cap R_i \neq 0$, since $_{R_i}(S_i \nabla R_i)$. Now $R_i x \cap R$ is a left R-submodule of $\Pi_i R_i$, and therefore $R \cap R_i x \supseteq J \cap (R_i x \cap R) \neq 0$. Since R is a subdirect sum of the family $\{R_i\}$, it is easy to see that $R \cap R_i$ and $R_i x \cap R$ are left ideals in R_i. Hence, primeness of R_i implies that $(R \cap R_i)(R_i x \cap R) \neq 0$. Again, since R is a subdirect sum of the $\{R_i\}$, we observe that $(R \cap R_i)R_i \subseteq (R \cap R_i)R$, so that

$$Rx \cap R \supseteq (R \cap R_i)(Rx \cap R) \supseteq (R \cap R_i)(R_i x \cap R) \neq 0.$$

Thus, $_R(S \nabla R)$, and so $S \cong \hat{R}$.

R, being a subdirect sum of prime rings, is semiprime. Since R has regular quotient ring S, $Z_1(R) = 0$ by Johnson's theorem [3].+) Thus, if Q denotes the injective hull of $_R R$, by the results stated at the beginning of this section, we have that

---

*) This is also given on p.73 of these Lectures, Theorem 8.

+) This theorem appears on p.70 of these Lectures, Proposition 3.

$S \cong \mathrm{Hom}_R(Q,Q)$. Then (1.15) implies that each nonzero direct summand of Q contains a m.d.s. We therefore can apply (2.4) (with R = M) to conclude that every nonzero r.c. left ideal of R contains a minimal r.c. left ideal. This completes the proof of the theorem.

The following corollary of Theorems 3.6 and 3.7 is immediate, since the left injectivity if R implies that $\hat{R} = R$, and since a direct product of full right linear rings is left self-injective.

COROLLARY 3.8. A ring $R \neq 0$ is a direct product of full right linear rings if and only if R satisfies the conditions: (a) R is semiprime; (b) R is left self-injective; and (3) each nonzero r.c. left ideal of R contains a minimal r.c. left ideal.

We also have the following characterization.

THEOREM 3.9. The following statements about a ring $R \neq 0$ are equivalent:

(I) R is semiprime, left self-injective, and each nonzero annihilator left ideal contains a minimal nonzero annihilator left ideal.

(II) R is a direct product of full right linear rings.

(III) R is semiprime, left self-injective, and each nonzero left ideal contains a minimal left ideal.

PROOF. The proof is cyclical.

(I) → (II). Let $\mathfrak{F}$ denote the totality of annihilator left ideals of R. Then $\mathfrak{F}$ satisfies conditions (a) - (c) of (3.1): (a) being satisfied by the definition of $\mathfrak{F}$; (b) being the last part of statement (I); and (c) is the evident statement that the intersection of two annihilator left ideals is an annihilator left ideal. Since it follows from (3.1) that $Z_1(R) = 0$, $\hat{R}$ is defined, and $\hat{R} = R$, since R is left self-injective. R is therefore a regular ring. By (2.4) the r.c. left ideals of R are the left direct summands of R, considered as a left R-module. Thus, if K is a r.c. left ideal, then $K = Re$, with $e = e^2 \in R$, so K is an annihilator left ideal. By (2.2), each annihilator left ideal is r.c., so a left ideal is r.c. if and only if it is an annihilator left ideal. Thus, each nonzero r.c. left ideal contains a minimal nonzero r.c. left ideal, so (I) → (II) follows from (3.8).

(II) ⇒ (III). Since each full right linear ring is left self-injective and regular, R has these properties. As we stated in the proof of (1.15), each nonzero left ideal of R contains a minimal left ideal, completing the proof.

(III) ⇒ (I). Assuming (III), each nonzero left ideal of R contains a minimal left ideal $P \neq 0$. Since R is semiprime, $P = Re$, for some idempotent $e \in R$ (Jacobson [2, p.57]). Thus, P is a minimal annihilator left ideal $\neq 0$, completing the proof.

We now connect our results with three results in the literature.

(i) Let R be a semiprime ring satisfying the minimum condition on annihilator left ideals. (Then each nonzero annihilator left ideal contains a minimal such). Then R satisfies the maximum condition on annihilator (2-sided) ideals. Then Levy's [Unique direct sums of prime rings, Trans. Amer. Math. Soc. 106 (1963) p.70, Theorem 3.13] states that R is an (irredundant) subdirect sum of a finite number of prime rings.

(ii) Let R be a semiprime ring satisfying the minimum condition on principal left ideals. (Then each nonzero left ideal of R contains a minimal left ideal.) In this case the structure of R is known (without assuming that R is left self-injective): R is a direct sum of simple rings $\{S_i\}$ each containing a minimal left ideal (= SMI rings). (Kaplansky [3], Faith [1].) Curiously enough, if we assume in addition that R contains an identity element, then each S_i does, and S_i is therefore a simple artinian ring $\forall i$. Then R is left and right self-injective (in fact, semisimple artinian) and a direct sum of finite dimensional full linear rings.

(iii) (1) of (3.5) is a special case of a known result (L.Levy - see above, p. 72, Prop. 4.2).

4. MATRIX SUBRINGS OF PRIME RINGS. If R is a prime ring which has a classical left quotient ring which is a full ring K^n of $n \times n$ matrices over a field K, then a theorem of Faith and Utumi [3] states that K is the left quotient field of a subring F such that R is isomorphic to a subring of K^n containing F^n. In this section we prove an analog of this result for a prime ring R whose maximal left quotient ring $\hat{R}$ is a full right linear ring. For brevity, such a ring R is said to be a prime

quotient-full ring.

Let K be a field, and n be either a positive integer or $+\infty$. We define an $n \times n$ matrix subsystem of K to be a collection $\{D_{ij}\}$ of nonzero additive subgroups of K (where $0 < i,j \leqq n$) satisfying the following conditions:

(a) $D_{ij}D_{jk} \subseteq D_{ik}$,

(b) D_{ij} is an essential left D_{ii}-submodule of K.

Observe that (b) makes sense, since it follows from (a) that D_{ii} is a subring of K and D_{ij} is a left D_{ii}-submodule of K.

We denote by K^n the ring of all row-finite $n \times n$ matrices over K, and by $M(D_{ij})$ the subset of K^n consisting of all matrices of the form

$$mI + (a_{ij}),$$

where m is an integer, I is the identity matrix, $a_{ij} \in D_{ij}$, and $a_{ij} = 0$ for almost all pairs $i,j (0 < i,j \leqq n)$. An easy computation shows that $M(D_{ij})$ is a subring of K^n, and the result below shows that it is a quotient-full prime ring.

(4.1) $M(D_{ij})$ is a prime ring (with zero left singular ideal), and K^n is its maximal left quotient ring.

PROOF. Let P, J be nonzero ideals of $R = M(D_{ij})$. Select nonzero elements $X = (x_{ij})$ in P and $Y = (y_{ij})$ in J. Then there exist indices i,j,k,l such that $x_{ij} \neq 0$, and $y_{kl} \neq 0$. Select a nonzero element $a_{1i} \in D_{1i}$ and a nonzero $b_{j1} \in D_{j1}$. Let A be the matrix with a_{1i} at the intersection of the first row and i-th column, with zeros elsewhere, and let B be the matrix with b_{j1} in the $(j,1)$-position and zeros elsewhere. Then $A, B \in R$, and $AXB \in P$ has $a_{1i}x_{ij}b_{j1}$ in $(1,1)$-position, and zeros elsewhere. Similarly, we can find elements $A', B' \in R$ such that $A'YB' \in J$ has a nonzero entry $a'_{1k}y_{kl}b'_{l1}$ in the $(1,1)$-position, and zeros elsewhere. Since K is a field, it is clear that $(AXB)(A'YB') \neq 0$, so $PJ \neq 0$. Hence R is a prime ring.

Since K^n is left self-injective, and is a regular ring, in order to show that $K^n = R$, we need only show that R is an essential left R-submodule of K^n, notationally ${}_R(K^n \nabla R)$. Let $X = (x_{ij})$ be a nonzero element of K^n. Then some row of X, say the i-th, contains nonzero elements. Let $x_{ij_1}, \ldots, x_{ij_t}$ be the nonzero elements of this row. (Remember that the matrices of K^n are row-finite.) Since $D_{ij_1},\ldots,D_{ij_t}$ are essential left D_{ii}-submodules of K, it is clear that there exists $a \in D_{ii}$ such that ax_{ij_k} is a nonzero element of D_{ij_k}, for $k = 1,\ldots,t$. Let A be the matrix with a in the (i,i)-position and zeros elsewhere. Then $A \in R$, and AX can be seen to be a nonzero element of R. Thus ${}_R(K^n \nabla R)$ as asserted, completing the proof.

In the remainder of this section R will be a prime quotient-full ring and S, a full right linear ring, will denote $\hat{R}$.

(4.2) Let e be a primitive idempotent in S. Then there exists a primitive idempotent e' in S such that $ee' \neq 0$ and $R \cap e'S \neq 0$.

PROOF. Select $b \in R$ such that $be \neq 0$ and $be \in R$. Then $J = R \cap beS$ is a nonzero right ideal of R. Also $P = R \cap Se$ is a nonzero left ideal of R. Since R is prime, $PRJ \neq 0$, so there exists $a \in R$ such that $eaJ \neq 0$. Then $abeS \supseteq aJ \neq 0$, and so $R \cap (abeS) \neq 0$. Since abeS is a nonzero epimorph of the minimal right S-module eS, abeS is a minimal right ideal of S, so there exists a primitive idempotent $e' \in S$ such that $abeS = e'S$. Then $R \cap e'S \neq 0$. Since $ee'S \supseteq eaJ \neq 0$, $ee' \neq 0$, completing the proof.

(4.3) Let $\{e_i\}$ be a collection of primitive idempotents of S. Then there exists a collection $\{f_i\}$ of primitive idempotents of S such that $Se_i = Sf_i$ and $R \cap f_iSf_j \neq 0, \forall i,j$.

PROOF. Given e_i, select e_i' satisfying the conditions of (4.2). Since S is semiprime, it is known that there exists a (primitive) idempotent $f_i \in S$ such that $Sf_i = Se_i$ and $e_i'S = f_iS$. Then $R \cap Sf_i \neq 0$, and $R \cap f_iS = R \cap e'S \neq 0$. Since R is prime, it follows that

$$0 \neq (R \cap f_iS)(R \cap Sf_j) \subseteq R \cap f_iSf_j \forall i,j,$$

completing the proof.

(4.4) Let e,e' be primitive idempotents in S, and let $D = R \cap eSe$, a subring of the field eSe. If $R \cap eSe'$ and $R \cap e'Se$ are both nonzero, then $R \cap eSe'$ is an essential left D-submodule of eSe'.

PROOF. Set $K = (R:x) \cap (R:e)$, where x is a given nonzero element of eSe'. Since K is an essential left ideal of R, and since $Z_1(R) = 0$, it follows that $Kx \neq 0$. Since the idempotents e and e' are primitive, and $R \cap eSe'$, $R \cap e'Se$ are nonzero, it follows that

$$J = Kx(R \cap e'Se)(R \cap eSe') \neq 0.$$

Primeness of R yields $J^2 \neq 0$, so $(R \cap eSe')Kx \neq 0$. Since $x = ex$, and since $Ke \subseteq R$, we obtain that $K' = (R \cap eS')Ke$ is a left ideal of D, and

$$R \cap eSe' \supseteq (R \cap eSe')Kx = K'x \neq 0.$$

This completes the proof.

(4.5) THEOREM. Let R be a prime ring with identity element and left quotient ring S, which is assumed to be a full right linear ring over a field K. Furthermore, assume: (*) S is also a (not necessarily maximal) right quotient ring of R. Then there exists an $n \times n$ matrix subsystem $\{D_{ij}\}$ of K (where n is a positive integer or $+\infty$) and a subring $\bar{R}$ of K^n containing $M(D_{ij})$ such that $R \cong \bar{R}$.

PROOF. As stated earlier, each nonzero left ideal of S contains a minimal left ideal, hence contains a primitive idempotent of S. Let $\{e_i\}$ denote a maximal set of orthogonal primitive idempotents of S. Then the sum

$$J = \sum_i Se_i$$

is direct and J is a two-sided ideal of S containing every minimal ideal. By (*) we can assume[1)] that $R \cap e_iSe_j \neq 0 \ \forall\, i,j$.

Now let e be a primitive idempotent of S. Since S is a full right linear ring over K, it is clear that $\mathrm{Hom}_S(Se,Se) \cong eSe \cong K$, and we shall identify these three fields. Furthermore, $Se \cong Se_i$ as left S-modules $\forall i$. Set $V = \mathrm{Hom}_S(Se,J)$. Then V

---

1) (*) implies that $K_i = e_iS \cap R \neq 0$. Since S is a left quotient ring of R, necessarily $Q_j = Se_j \cap R \neq 0$. Then, since R is prime, and since K_i is a right ideal of R, necessarily $K_{ij} = K_iQ_j \neq 0$. Clearly $K_{ij} \subseteq e_iSe_j \cap R$, proving the assertion.

is a (K,S)-bimodule, and $S \cong Hom_K(V,V)$ by (1.11). For each i let y_i be an isomorphism of Se onto Se_i ; then $y_1, y_2, \ldots$ is a K-basis of V which defines in the usual way an isomorphism $\sigma : S \cong K^n$, where $n = \dim V$ is a positive integer or $+\infty$. Observe furthermore that $Ky_i = Hom_S(Se, Se_i)$.

Let $a \in R \cap e_iSe_j$. Then clearly $y_i a \in Hom_S(Se, Se_j) = K_{y_j}$, and so there is a unique element of K, which we denote by $\sigma_{ij}(a)$, such that $y_i a = \sigma_{ij}(a) y_j$. Thus for each pair i,j we obtain a mapping $\sigma_{ij} : R \cap e_iSe_j \to K$, which is clearly 1-1. We omit the trivial verification that, if $a_{ij} \in R \cap e_iSe_j$, and $a_{jk} \in R \cap e_jSe_k$, then $a_{ij}a_{jk} \in R \cap e_iSe_k$ and $\sigma_{ik}(a_{ij}a_{jk}) = \sigma_{ij}(a_{ij})\sigma_{jk}(a_{jk})$. Let D_{ij} be the image of $R \cap e_iSe_j$ in K under the mapping σ_{ij} . We then obtain from the above discussion, and from (4.4), that $\{D_{ij}\}$ is an $n \times n$ matrix subsystem of K.

Now let R_o be the additive subgroup of R generated by 1 and $R \cap e_iSe_j \forall i,j$; R_o is clearly a subring of R. If $a \in R \cap e_iSe_j$, then since $y_i a = \sigma_{ij}(a) y_i \forall i,j$ and $y_k a = 0$ for $k \neq i$, it follows that $\sigma(a)$ is simply the matrix with $\sigma_{ij}(a)$ in the (i,j)-position and zeros elsewhere. From this it follows that $\sigma(R_o) = M(D_{ij})$. Setting $\bar{R} = \sigma(R)$, we see that $R \cong \bar{R}$ and $M(D_{ij}) \subseteq \bar{R} \subseteq K^n$.

REMARKS 1. (4.1) shows that K^n is also the maximal left quotient ring of $\bar{R}$.

2. The assumption (*) can be replaced by the assumption that S contains a maximal set of orthogonal primitive idempotents $\{e_i\}$ having the property that $e_iSe_j \cap R \neq 0$ for all i,j. By Footnote 1 it is enough to assume that $e_iS \cap R \neq 0$ for all i. It is true that (4.3) implies that there exists a corresponding collection of primitive idempotents $\{f_i\}$ such that $f_iSf_j \cap R \neq 0$ for all i,j, but we have not seen how to do this in a way which would guarantee the orthogonality of the $\{f_i\}$, and the proof of (4.5) depends on the orthogonality of the $\{e_i\}$.

14. JOHNSON RINGS

Consider a quotient-simple ring R whose quotient ring is $S \cong D_n$, where D is a division ring, and represent S as a full ring of linear transformations on a left vector space V over D. If B is a basis of V, let $\dot{d}$ denote the scalar l.t in S defined relative to B by $d \in D$. Then, (11.6) implies that D contains an Ore domain F satisfying $\hat{F} = D$ such that if $s \in S$, then there exists $0 \neq k \in F$, $r \in R$ such that $s = (\dot{k})^{-1}a$. In particular $S = \{\dot{d}r \mid d \in D,\ r \in R\}$. It is this aspect of (11.6) which is generalized in this section.

A right ideal I of a ring R is <u>closed</u> (§7) in case I_R has no essential extension in R_R other than I_R . By Zorn's lemma, each right ideal is contained in a right ideal which is both an essential extension and closed. If the right singular ideal $Z_r(R) = 0$, then (§8) each right ideal I is contained in a unique maximal essential extension I' (which is obviously closed), and the mapping $I \to I'$ has the following properties:

(P1) $I' \supseteq I$ (P2) $(I')' = I'$

(P3) $I \supseteq J \to I' \supseteq J'$ (P4) $0' = 0$

(P5) $I \cap J = 0 \to I' \cap J' = 0$

(P6) $aI' \subseteq (aI)'$ $\forall\, a \in R$

where I, J are arbitrary right ideals of R.

DEFINITION. A right ideal I of R is prime if the following implication holds for any two right ideals A and B of R: $AB \subseteq I$ and $B \neq 0 \to A \subseteq I$.

A prime ring A has a <u>right structure</u> (Johnson [2]) in case there exists a mapping $I \to I^*$ of the set of right ideals into the set P of prime right ideals of R such that (P1) - (P6) is satisfied (with I^* playing the role of I'), and in addition the following condition holds: (P7) P contains minimal elements $\neq 0$. Let $(A,*)$ denote that A has a right structure.

1. PROPOSITION. A prime ring R has a right structure if and only if the following two conditions hold: (1) $Z_r(R) = 0$; (2) R has a minimal closed right ideal.

Then $I \to I'$ defines a unique right structure $(R,')$ for R.

PROOF. We already have remarked that (1) implies that $I \to I'$ satisfies (P1)-(P6). If R is prime, then the closed right ideals of R are prime right ideals. Then (2) implies that R has the right structure $(R,')$.

Conversely, if R has a right structure $(R,*)$, then the set $\{I^*\}$ coincides with the totality of closed (= complement) right ideals of R (Johnson [1, pp. 805-6]). Then (2) holds. Furthermore, if $a \in Z_r(R)$, then $aI = 0$ for some large right ideal of R, and since $I^* = R$, applying (P4) and (P6) we have:

$$0 = 0^* = (aI)^* \supseteq aI^* = aR .$$

Thus $aR = 0$, and primeness of R implies $a = 0$. Hence (1) holds, proving the proposition.

We consider a class of prime rings having a left and right structure, and call these Johnson rings. By the proposition we have:

DEFINITION. A Johnson ring is a prime ring satisfying $Z_r(R) = Z_l(R) = 0$, and containing a minimal closed right ideal U and a minimal closed left ideal W.

TRANSITIVITY THEOREM (Johnson [2, Theorem 3.3]). Let R be a Johnson ring as above. Then: (1) it can be assumed that $WU \neq 0$; (2) then $K = U \cap W$ is an Ore domain and U is a torsion-free left K-module; (3) if $x_1,\dots,x_n$ are finitely many K-linearly independent elements of U, and if $y_1,\dots,y_n \in U$, then there exist $0 \neq k \in K$, $r \in R$ such that $x_i r = ky_i$, $i = 1,\dots,n$.

From now on, the Johnson ring R will be endowed with the properties and notations of the transitivity theorem.

2. LEMMA. U is embedded in left vector space V over the quotient division ring D of K with the following properties: (1) $V = \{k^{-1}x \mid 0 \neq K \in K,\ x \in U\}$; (2) If $x_1,\dots,x_n \in V$, then there exists $0 \neq q \in K$ such that $qx_i \in U$, $i = 1,\dots,n$; (3) any maximal K-linearly independent subset of U is a basis of V over D.

PROOF. Since U is a left K-module, and D is a right K-module, the tensor product $V = D\otimes_K U$ exists. Since ${}_KU$ is torsion-free, V is a vector space over D under the scalar operation: $c(d \otimes x) = (cd) \otimes x$ $d, c \in D$, $x \in U$. Obviously, U is embedded in V under the correspondence $x \to 1 \otimes x \;\forall\; x \in U$. If $d = k^{-1}q \in D$, then $d \otimes x = k^{-1}(q \otimes x) = k^{-1}(1 \otimes qx)$, establishing (1). (2) follows from the common left multiple property of K, since if $x_i = k_i^{-1}y_i$, $y_i \in U$, $i = 1,\dots,n$, then there exists $0 \neq q \in \bigcap_{i=1}^{n} Kk_i$, $q = h_ik_i$, $h_i \in K$, and then $qx_i = h_iy_i$ U, $i = 1,\dots,n$. (3) is equally trivial.

3. THEOREM. Let R be a Johnson ring, and let V be the vector space of the lemma. Then R can be considered as a subring of $A = \mathrm{Hom}_D(V,V)$. (1) If $x_1,\dots,x_n$ are finitely many linearly independent vectors of V, and if $y_1,\dots y_n \in V$, then there exist $0 \neq d \in D$, $r \in R$ such that $x_ir = dy_i$, $i = 1,\dots,n$. (2) If $\dot{d}$ denotes the scalar l.t. relative to a given basis B of V defined by $d \in D$, then the subset $S = \{\dot{d}r \mid d \in D, r \in R\}$ is dense in the finite topology on A.

PROOF. First let $\{u_\lambda \mid \lambda \in \Lambda\}$ be a basis of V consisting of elements of U. If $r \in R$, then $y_\lambda = u_\lambda r \in U$ $\forall \lambda \in \Lambda$, and the correspondence $u_\lambda \to y_\lambda$ defines an element $\bar{r} \in A$. Obviously, $\bar{R} = \{\bar{r} \in A \mid r \in R\}$ is a subring of A, and $r \to \bar{r}$ is a homomorphism of R onto $\bar{R}$. If $\bar{r} = 0$, then $U\bar{r} = 0$, so $Ur = 0$, and then $r = 0$ by primeness of R. Since $R \cong \bar{R}$, we can identify R with $\bar{R}$.

(1) By 2, there exists $0 \neq q \in K$ such that qx_i, $qy_i \in U$, $i = 1,\dots,n$. Since $qx_i,\dots,qx_n$ are K-linearly independent, by Johnson's transitivity theorem there exist $0 \neq k \in K$, $r \in R$, such that $(qx_i)r = k(qy_i)$, $i = 1,\dots,n$. Then, $x_ir = dy_i$, $i = 1,\dots,n$, where $d = q^{-1}kq \in D$.

(2) Let $B = \{v_\lambda \mid \lambda \in \Lambda\}$ be a basis of V. There exists a finite subset $v_1,\dots,v_t \in B$ such that

$$x_i = \Sigma_{j=1}^{t} d_{ij}v_j$$

with $d_{ij} \in D$, $i = 1,\dots,n$, $j = 1,\dots,t$. Let $a \in A$ be such that $x_ia = y_i$, $i = 1,\dots,n$. By (1), there exist $0 \neq d \in D$, $r \in R$ such that

$$v_jr = d^{-1}(v_ja), \qquad j = 1,\dots,t.$$

Then

$$v_j(\dot{d}r) = v_j a \ , \qquad j = 1,\ldots,t,$$

and

$$x_i(\dot{d}r) = \sum_{\substack{i=1,\ldots,n \\ j=1,\ldots,t}} d_{ij}(v_j\dot{d}r) = \sum_{i,j} d_{ij}(v_j a) = x_i a = y_i \ ,$$

$i = 1,\ldots,n$. Thus, the subset $S = \{\dot{d}r \mid d \in D,\ r \in R\}$ is dense in the finite topology on A.

REMARK. Let R be a prime ring with the following properties:

(1) $Z_l(R) = 0$

(2) The maximal left quotient ring S of R is also a right quotient ring of R;

(3) S contains a minimal left ideal I.

Then (2) implies that $Z_r(R) = 0$, and it follows from the Lattice isomorphisms $L_l(S) \cong L_l(R)$, $L_r(S) \cong L_r(R)$, that $W = I \cap R$ (resp. $U = J \cap R$, where J is a minimal right ideal of S) is a minimal closed left (resp. right) ideal of R. Thus, R *is a Johnson ring*.

15. OPEN PROBLEMS

All rings are assumed to have an identity element 1.

1. Kaplansky has shown that a commutative ring is regular if and only if: (*) each simple right R-module is injective. (See Rosenberg-Zelinsky [1]). Recently Villamayor has shown for an arbitrary ring R that (*) is equivalent to the requirement that each right ideal of R is the intersection of maximal right ideals (unpublished).

   QUESTION. Does (*) imply that R is regular? (It is known that regularity of R does not imply (*).) Cf. Problem 17 below.

2. Characterize the endomorphism rings of quasi-injective modules; of quasi-injective modules of finite length.

3. Let M_R be a module of finite length satisfying the double annihilator relation $S^{RM} = S$ (see §3) for finitely generated Λ-submodules S, where $\Lambda = \mathrm{Hom}_R(M,M)$. What can be said about the structure of M_R, e.g., is M_R necessarily quasi-injective?

4. (Matlis [1]). Let (P) denote the property: A module is a direct sum of minimal injective modules. (Every INJECTIVE module over a right Noetherian ring has property (P).) If M_R has property (P), does each direct summand of M have property (P)?*)

5. If R is right self-injective ring with identity, then Utumi's theorem states that J(R) (= Jacobson radical) coincides with $Z_r(R)$ (= the right singular ideal.) Does there exist such a situation for which $J(R) = Z_r(R)$ is not a nil ideal? (Conjecture: yes; Cf. Faith and Utumi [2].)+)

6. If R is a ring such that R/I is injective for each right ideal I (= every cyclic R-module is injective), then R is a semisimple Artinian (Osofsky [1]). Characterize R having the property that R/I is injective whenever I is a

---

*) This has been solved under the assumption that M is injective by Faith and Walker (See the Added Bibliography.)

+) F.W. Anderson saw through this one right away. Let R be the direct product of the rings Z/p^nZ, one for each prime p. Then R is self-injective with nonnil radical.

finitely generated right ideal. (Conjecture: R is an arbitrary right self-injective regular ring.)

7. Does there exist a primitive ring R such that the singular ideal $Z_r(R) \neq 0$? If so, then $S = Z_r(R)$ is a primitive ring, and $Z_r(S) = S$. Barbara Osofsky [3] has shown that there exists a semiprimitive ring R, i.e., with radical of $R = 0$, such that $Z_r(R) \neq 0$, but she has not been able to determine whether or not R in her example is primitive.

8. (Goldie). Let R be a right noetherian prime ring, and let $R \supset P_1 \supset \dots \supset P_n \supset \dots$ be a sequence of prime ideals. Then, for each k, R/P_k is a rt. noetherian prime ring, hence by Goldie's Theorem, has a classical right quotient ring which is a ring of $m_k \times m_k$ matrices over some field D^k.

QUESTIONS: (1) How does the sequence $m_1, m_2, \dots, m_k, \dots$ behave? Is it bounded? If $m_1 = m_2 = \dots = m_k = \dots$, how are the fields D^k related?

9. (Jacobson). If R is right quotient-simple, is each Jordan homomorphism of R a sum of a homomorphism and an anti-homomorphism? (Cf. Jacobson and Rickart, Jordan homomorphisms of rings, Trans. Amer. Math. Soc. 69 (1950) 479-502.)

10. A ring S is right intrinsic over a ring R in case each non-zero right ideal of S has non-zero intersection with R. Characterize the right intrinsic extensions of R in case R is (a) right quotient-simple, or right quotient-simple; (b) $Z_r(R) = 0$. (Cf. Faith and Utumi, Intrinsic extensions of rings, Pacific J. Math. 1964).

11. Let U be a torsion-free left module over an integral domain K having a left quotient field D (K is a left Ore domain). Then $V = D \otimes_K U$ is a left vector space over D.

QUESTION: What is the relationship between $R = \mathrm{Hom}_K(U,U)$ and $S = \mathrm{Hom}_D(V,V)$? (Note that R is embedded in S under the map $r \to \bar{r}$, where if $r \in R$, then $(k^{-1} \otimes x)\bar{r} = k^{-1} \otimes xr$, for all $0 \neq k \in K$, $x \in U$.) In the case U is a finitely generated over K, and K is a left and right Ore domain, then Feller and Swokowski have shown that S is a classical left and right quotient-ring of R. Is S the maximal right quotient ring of R in every case, i.e., even if U is not finitely generated, or even if K is just a left Ore domain?*)

12. Stephen U. Chase has constructed an example of a commutative integral domain K having the property that each countable collection of non-zero ideals has non-zero intersection (unpublished). From this it follows that the ring D_∞ of all row-finite matrices over the quotient field D of K, is a classical right (and left) quotient ring of K_∞. Expressed otherwise, if W is a free K-module on a countable set, and if V is the D-vector space $D\otimes_K W$, then $D_\infty = \mathrm{Hom}_D(V,V)$ is a classical quotient-ring of $K_\infty = \mathrm{Hom}_k(W,W)$. Thus, it is legitimate to ask: Characterize the rings R having a classical right quotient ring S of the form $S = \mathrm{Hom}_D(V,V)$, where V is a right vector space over a field D.+) (Goldie [2] did this for the case $[V:D] < \infty$.)

13. Let R be a ring with singular ideal $Z_r(R) = 0$, such its maximal right quotient ring $\hat{R}$ is a ring D_n of $n \times n$ matrices over a field D. Does there exist some choice of matrix units in D_n, and a right order K in D, such that R contains all first column matrices $Ke_{11} + \dots + Ke_{n1}$ with coefficients in K? (Cf. Faith-Utumi [3]).

14. Does there exist a one-sided version of Johnson's transitivity theorem, as in the case of right primitive rings? If R is a prime ring with $Z_r(R) = 0$, and containing a minimal closed right ideal U, then $K = \mathrm{Hom}_R(U,U)$ is a right Ore domain. If $x_1,\dots,x_n$ are left K-linearly independent elements of U, and if $y_1,\dots,y_n \in U$, does there exist $0 \neq k \in K$, and $r \in R$, such that $x_i r = k y_i$, $i = 1,\dots,n$? (Added 1967: Koh and Mewborn [see added bibliography] have something on this problem, but the coefficients do not lie in K, rather in the quotient field of K.)

15. (R.E. Johnson). Let K be an integral domain, not an Ore domain. Thus, K does not satisfy the a.c.c. on complement right ideals nor on complement left ideals.

---

*) J. Zelmanowith has given fairly extensive answers to this question. See his forthcoming paper listed in the Added Bibliography. See also R. Hart's paper in the same journal.

+) This appears to be a difficult problem, but, along with Goldie's problem 8, listed above, it is the most interesting one.

Let n be an integer > 1. Does the matrix ring K_n satisfy the a.c.c. on annihilator right ideals?

16. Is a right self-injective simple ring necessarily left self-injective? Recall that if K is a simple integral domain, then the maximal quotient ring Q of K is a right self-injective regular simple ring. Is Q also left self-injective?

17. Addition to Problem 1 stated above. The following two conditions are equivalent:

(1) Every simple right R-module is injective.

(2) Every right ideal of R is the intersection of maximal right ideals.

F.W. Anderson told us that this theorem belongs to Villamayor. When R is commutative, (2) is equivalent to

(3): R is a regular ring.

In this case, the equivalence of (1) and (3) is a theorem of Kaplansky. It is known that a noncommutative regular ring need not satisfy (1). The converse seems to be unknown.

Call a ring R a V-ring if it satisfies the equivalent conditions (1) and (2).

THEOREM. *If R is a prime ring with ascending chain conditions complement right ideals and annihilator right ideals, and if R is a V-ring, then R is a simple ring.*

PROOF. Let I be an ideal $\neq 0$. If $I \neq R$, then I is contained in a maximal right ideal M. By Goldie's theorem, I contains a regular element x of R. Since $V = R/M$ is injective, V is divisible by every regular element of R (cf. § 1). But $x \in M$, so $Vx = V = 0$, a contradiction.

A. Ornstein has remarked that the proof shows that any integral domain which is a V-ring is a simple ring. In his thesis [see added bibliography], Ornstein shows that every semiprime noetherian V-ring is a direct sum of simple V-rings.

Question (1): *Does there exist a simple noetherian V-ring which is not semisimple?*

(2): *Is every V-ring a regular ring?*

An affirmative answer to (2) implies that to (1). A regular, however, need not be a V-ring. Such an example is the ring $S + F$, where S is the socle of

$End_F V$, where V is an infinite vector space over a field F. (Here $S + F$ denotes socle plus all scalars. My proof of this example uses the theory of quotient rings.)

REFERENCES

E. Artin

[1] The influence of J.H.M. Wedderburn on the development of modern algebra. Bull. Amer. Math. Soc. 56 (1950) 65-72.

[2] Rings with minimum condition. University of Michigan Press, 1940, Ann Arbor, Michigan.

R. Baer

[1] Abelian groups which are direct summands of every containing group. Proc. Amer. Math. Soc. 46 (1940) 800-806.

H. Bass

[1] Global dimensions of rings, Ph.D. Thesis. University of Chicago, 1959.

[2] Finitistic homological dimension and a homological generalization of semiprimary rings. Trans. Amer. Math. Soc. 95 (1960) 466-488.

H. Cartan and S. Eilenberg

[1] Homological algebra. Princeton University Press, 1956, Princeton, N.J.

B. Eckmann and A. Schopf

[1] Ueber injektive Moduln. Archiv der Math. 4 (1953) 75-78.

C. Faith and Y. Utumi

[1] Quasi-injective modules and their endomorphism rings. Archiv der Math. 15 (1964), 166-174.

[2] Baer Modules, Ibid.

[3] On noetherian prime rings. Trans. Amer. Math. Soc. 114 (1965), 53-60.

[4] Intrinsic extensions of rings. Pac. J. Math. 14 (1964).

C. Faith

[1] Rings with minimum condition on principal ideals I;II. Archiv der Math. 10 (1959) 327-330; 12 (1961) 179-181.

[2] Orders in simple artinian rings. Trans. Amer. Math. Soc. 114 (1964), 61-65.

G. D. Findlay and J. Lambek

[1] A generalized ring of quotients I,II. Can. Math. Bull. 1 (1958) 77-85, 155-167.

A. W. Goldie

[1] The structure of prime rings under ascending chain conditions. Proc. Lond. Math. Soc. 8 (1958) 589-608.

[2] Semiprime rings with maximum condition. Proc. Lond. Math. Soc. 10 (1960) 201-220.

[3] Non-commutative principal ideal rings. 13 (1962) 213 ff.

C. Hopkins

[1] Rings with minimal conditions for left ideals. Ann. of Math. 40 (1939) 712-730.

N. Jacobson

[1] Theory of rings. Amer. Math. Soc. Surveys, Vol. 2, 1943, New York.

[2] Structure of rings. Amer. Math. Soc. Colloquium, Vol. 36, revised edition, 1964, Providence, R.I.

R. E. Johnson

[1] Prime rings. Duke Math. J. 18 (1951) 799-809.

[2] Representations of prime rings. Trans. Amer. Math. Soc. 74 (1953) 351-357.

[3] Extended centralizer of a ring over a module. Proc. Amer. Soc. 2 (1951) 891-895.

[4] Quotient rings of rings with zero singular ideal. Pac. J. Math. 11 (1961) 1385-1392.

R. E. Johnson and E. T. Wong

[1] Quasi-injective modules and irreducible rings. J. Long. Math. Soc. 36 (1961) 260-268.

[2] Self-injective rings. Can. Math. Bull. 2 (1959) 167-173.

I. Kaplansky

[1] Modules over dedekind rings and valuation rings. Trans. Amer. Math. 72 (1952) 327-340.

[2] Projective modules. Annals of Math. 68 (1958) 372-377.

[3] Topological representations of algebras II. Trans. Amer. Math. Soc. 68 (1950) 62-75.

J. Lambek

[1] On the structure of semiprime rings and their rings of quotients. Can. J. Math. 13 (1961) 392-417.

[2] On Utumi's ring of quotients. Can. J. Math. 15 (1963) 363-370.

L. Lesieur and R. Croisot

[1] Sur les anneaux premier noetherien à gauche. Ann. Sci. Ecole Norm. Sup. 76 (1959) 161-183.

J. Levitzki

[1] Prime ideals and the lower radical. Amer. J. Math. 73 (1951) 23-29.

[2] Solution of a problem of G. Koethe. Amer. J. Math. 67 (1945) 437-442.

[3] On rings which satisfy the minimum condition for right-hand ideals. Compositio Math. 7 (1939) 214-222.

E. Matlis

[1] Injective modules over noetherian rings. Pac. J. Math. 8 (1958) 511-528.

O. Ore

[1] Linear equations in non-commutative fields. Ann. of Math. 32 (1931) 463-477.

[2] Theory of non-commutative polynomials. Ann. of Math. 34 (1933) 480-508.

B. Osofsky

[1] Rings all of whose finitely generated modules are injective. Pac. J. Math. 14 (1964).

[2] On ring properties of injective hulls. Can. Math. Bull. 7 (1964).

[3] A semiprimitive ring with nonzero singular ideal. Notices Amer. Math. Soc. 10 (1963) Abstract 63T-173, p. 357.

A. Rosenberg and D. Zelinsky

[1] On the finiteness of the injective hull. Math. Z. 70 (1959) 372-380.

Y. Utumi

[1] On quotient rings. Osaka Math. J. 8 (1956) 1-18.

[2] On a theorem on modular lattices. Proc. Japan Acad. 35 (1959) 16-21.

[3] On rings of which any one-sided quotient ring are two-sided. Proc. Amer. Math. Soc. 14 (1963) 141-147.

ADDED BIBLIOGRAPHY

Bergman, G., A ring primitive on the right but not on the left, Proc. Amer. Math. Soc. 15 (1964), 473-475, Erratum.

Caldwell, W., Hypercyclic rings, Ph.D. Thesis, Rutgers U., 1966.

Faith, C., Algebra: Rings and Modules, W. B. Saunders, Philadelphia, 1968.

_________, Rings with ascending condition on annihilators, Nagoya Math. J. 27-1 (1966), 179-191.

_________, A general Wedderburn theorem, Bull. Amer. Math. Soc. 73 (1967), 65-67.

Faith, C., and Walker, E.A., Direct sum representations of injective modules, J. Algebra 5 (1967), 203-221.

Feller, E. H., and Swokowski, E. W., Reflective N-prime rings with the ascending chain condition, Trans. Amer. Math. Soc. 99 (1961), 264-271.

_____________________________, The ring of endomorphisms of a torsionfree module. J. London Math. Soc. 39 (1964), 41-42.

Gabriel, P., Des Catégories Abéliennes, Bull. Soc. Math., France, 90 (1962), 323-448.

Goldie, A. W., Non-commutative principal ideal rings, Archiv der Math. 13 (1962), 213-221.

____________, Torsion-free modules and rings, J. Algebra (1964), 268-287.

____________, Localization in non-commutative noetherian rings, J. Algebra 5 (1967), 89-105.

____________, A note on non-commutative localization.

R.N. Gupta and F. Saha, Remarks on a paper of Small, J. Math. Sci. 2 (1967), 7-16.

Harada, M., Quasiinjective modules, Osaka Math. J. 2 (1965), 351-356.

Hart, R., Simple rings with uniform right ideals, Proc. Lond. Math. Soc.

________, Endomorphisms of modules over semi-prime rings, J. Algebra 4 (1966) 46-51.

Herstein, I. N., A counterexample in noetherian rings, Proc. Nat. Acad. (USA) 54 (1965), 1035.

_____________, Theory of rings, University of Chicago mimeographed notes, 1965.

Herstein, I. N., and L. Small, Nil rings satisfying certain chain conditions, Can. J. Math. 11 (1962), 180-184. Addendum, ibid, 14 (1965), 300-302.

Jaffard, P., Les Systèmes d'Ideaux, Dunod, Paris, 1960.

Jans, J., On orders in QF rings, University of Washington, Seattle, 1966. (dittoed)

_______, A note on injectives, ibid. (submitted to Math. Ann.)

Johnson, R. E., Distinguished rings of linear transformations, Trans. Amer. Math. Soc. 111 (1964), 400-412.

_______, Potent rings, Trans. Amer. Math. Soc. 119(1965) 524-534.

_______, Prime matrix rings, Proc. Amer. Math. Soc. 16 (1965), 1099-1105.

_______, Rings of finite rank, Publicationes Math.(Debrecen) 11 (1964), 284-287.

_______, Principal right ideal rings, Canadian J. Math. 15 (1963), 297-301.

_______, Rings with zero right and left singular ideals,Trans.Amer.Math.Soc. 118 (1965), 150-157.

Koh and Newborn, Prime rings with maximal annihilator and maximal complement right ideals, Proc. Amer. Math. Soc. 16 (1965), 1073-1076.

_______, The weak radical of a ring, Proc. Amer. Math. Soc. 18 (1967), 554-559.

Lambek, J., Lectures on rings and modules, Blaisdell, Toronto, London, and Waltham (Mass.), 1966.

_______, On the ring of quotients of a noetherian ring., Can. Math. Bull. 8 (1965), 279-290.

Levy, L., Torsionfree and divisible modules over non-integral domains, Can. J. Math. 15 (1963), 132-151.

_______, Unique direct sums of prime rings, Trans. Amer. Math. Soc. 106 (1963), 64-76.

Ornstein, A., Rings with restricted minimum condition, Ph.D. Thesis, Rutgers, 1966. (Also submitted to Proc. Amer. Math. Soc.)

Osofsky, B., Endomorphism rings of quasi-injective modules, Can. J. Math.

_______, A generalization of quasifrobenius rings, J. Algebra, 3 (1966), 373-386.

S. Singh and S.K. Jain, On pseudo injective modules and self pseudo injective rings, J. Math. Sci. 2 (1967), 23-31.

Small, L., Artinian quotient rings, J. Algebra 4 (1966), 13-41, Corrections, Ibid p. 505-507.

_________, An example in noetherian rings, Proc. Nat. Acad. (USA) 54 (1965), 1035-1036.

Talentyre, Quotient rings with maximum condition on right ideals, J. London Math. Soc. 38 (1963), 439-450.

_________, Quotient rings with minimum condition on right ideals, J. London Math. Soc. 41 (1966), 141-144.

Utumi, Y., On continuity and self-injectivity of a complete regular ring, Can. J. 18 (1966), 404-412.

_________, On continuous rings and self-injective rings, Trans. Amer. Math. Soc. 118 (1965), 158-173.

_________, Prime J-rings with uniform one-sided ideals, Amer. J. Math. 85 (1963), 583-596.

_________, Self-injective rings, J. Algebra 6 (1967), 56-64.

Walker, C.P., and E.A., Quotient categories and rings of quotients, Trans. Amer. Math. Soc.

Zelmanowitz, J., Endomorphismrings of torsionless modules, J. Algebra 5 (1967), 325-41.

Wu, and Jans, Quasiprojective modules, Illinois J. Math. 11 (1967), 439-448.

INDEX

A

B

C

D

E

F

I

M

P

Q

Offsetdruck: Julius Beltz, Weinheim/Bergstr.

Lecture Notes in Mathematics

Bisher erschienen/Already published

Vol. 1: J. Wermer, Seminar über Funktionen-Algebren. IV, 30 Seiten. 1964. DM 3,80

Vol. 2: A. Borel, Cohomologie des espaces localement compacts d'après J. Leray. IV, 93 pages. 1964. DM 9,–

Vol. 3: J. F. Adams, Stable Homotopy Theory. 2nd. revised edition. IV, 78 pages. 1966. DM 7,80

Vol. 4: M. Arkowitz and C. R. Curjel, Groups of Homotopy Classes. 2nd. revised edition. IV, 36 pages. 1967. DM 4,80

Vol. 5: J.-P. Serre, Cohomologie Galoisienne. Troisième édition. VIII, 214 pages. 1965. DM 18,–

Vol. 6: H. Hermes, Eine Termlogik mit Auswahloperator. IV, 42 Seiten. 1965. DM 5,80

Vol. 7: Ph. Tondeur, Introduction to Lie Groups and Transformation Groups. VIII, 176 pages. 1965. DM 13,50

Vol. 8: G. Fichera, Linear Elliptic Differential Systems and Eigenvalue Problems. IV, 176 pages. 1965. DM 13.50

Vol. 9: P. L. Ivănescu, Pseudo-Boolean Programming and Applications. IV, 50 pages. 1965. DM 4,80

Vol. 10: H. Lüneburg, Die Suzukigruppen und ihre Geometrien. VI, 111 Seiten. 1965. DM 8,–

Vol. 11: J.-P. Serre, Algèbre Locale. Multiplicités. Rédigé par P. Gabriel. Seconde édition. VIII, 192 pages. 1965. DM 12,–

Vol. 12: A. Dold, Halbexakte Homotopiefunktoren. II, 157 Seiten. 1966. DM 12,–

Vol. 13: E. Thomas, Seminar on Fiber Spaces. IV, 45 pages. 1966. DM 4,80

Vol. 14: H. Werner, Vorlesung über Approximationstheorie. IV, 184 Seiten und 12 Seiten Anhang. 1966. DM 14,–

Vol. 15: F. Oort, Commutative Group Schemes. VI, 133 pages. 1966. DM 9,80

Vol. 16: J. Pfanzagl and W. Pierlo, Compact Systems of Sets. IV, 48 pages. 1966. DM 5,80

Vol. 17: C. Müller, Spherical Harmonics. IV, 46 pages. 1966. DM 5,–

Vol. 18: H.-B. Brinkmann und D. Puppe, Kategorien und Funktoren. XII, 107 Seiten. 1966. DM 8,–

Vol. 19: G. Stolzenberg, Volumes, Limits and Extensions of Analytic Varieties. IV, 45 pages. 1966. DM 5,40

Vol. 20: R. Hartshorne, Residues and Duality. VIII, 423 pages. 1966. DM 20,–

Vol. 21: Seminar on Complex Multiplication. By A. Borel, S. Chowla, C. S. Herz, K. Iwasawa, J.-P. Serre. IV, 102 pages. 1966. DM 8,–

Vol. 22: H. Bauer, Harmonische Räume und ihre Potentialtheorie. IV, 175 Seiten. 1966. DM 14,–

Vol. 23: P. L. Ivănescu and S. Rudeanu, Pseudo-Boolean Methods for Bivalent Programming. 120 pages. 1966. DM 10,–

Vol. 24: J. Lambek, Completions of Categories. IV, 69 pages. 1966. DM 6,80

Vol. 25: R. Narasimhan, Introduction to the Theory of Analytic Spaces. IV, 143 pages. 1966. DM 10,–

Vol. 26: P.-A. Meyer, Processus de Markov. IV, 190 pages. 1967. DM 15,–

Vol. 27: H. P. Künzi und S.T. Tan, Lineare Optimierung großer Systeme. VI, 121 Seiten. 1966. DM 12,–

Vol. 28: P. E. Conner and E. E. Floyd, The Relation of Cobordism to K-Theories. VIII, 112 pages. 1966. DM 9.80

Vol. 29: K. Chandrasekharan, Einführung in die Analytische Zahlentheorie. VI, 199 Seiten. 1966. DM 16.80

Vol. 30: A. Frölicher and W. Bucher, Calculus in Vector Spaces without Norm. X, 146 pages. 1966. DM 12,–

Bitte wenden/Continued